Erode G. Mahadevan

Ammonium Nitrate Explosives for Civil Applications

Related Titles

Koch, Ernst-Christian

**Metal-Fluorocarbon Based
Energetic Materials**

2012
ISBN: 978-3-527-32920-5

Lackner, M., Winter, F., Agarwal, A. K.
(eds.)

Handbook of Combustion

2010
ISBN: 978-3-527-32449-1

Agrawal, J. P.

High Energy Materials

Propellants, Explosives and Pyrotechnics

2010
ISBN: 978-3-527-32610-5

Meyer, R., Köhler, J., Homburg, A.

Explosives

2007
ISBN: 978-3-527-31656-4

Agrawal, J. P., Hodgson, R.

**Organic Chemistry of
Explosives**

2007
ISBN: 978-0-470-02967-1

Kubota, N.

Propellants and Explosives

Thermochemical Aspects of Combustion

2007
ISBN: 978-3-527-31424-9

Teipel, U. (ed.)

Energetic Materials

Particle Processing and Characterization

2005
ISBN: 978-3-527-30240-6

Erode G. Mahadevan

Ammonium Nitrate Explosives for Civil Applications

Slurries, Emulsions and Ammonium Nitrate Fuel Oils

WILEY-VCH

WILEY-VCH Verlag GmbH & Co. KGaA

The Author

Dr. Erode G. Mahadevan
Technology Consultant
C-22 Vikrampuri Colony
Secunderabad 500009
India

■ All books published by **Wiley-VCH** are carefully produced. Nevertheless, authors, editors, and publisher do not warrant the information contained in these books, including this book, to be free of errors. Readers are advised to keep in mind that statements, data, illustrations, procedural details or other items may inadvertently be inaccurate.

Library of Congress Card No.: applied for

British Library Cataloguing-in-Publication Data
A catalogue record for this book is available from the British Library.

Bibliographic information published by the Deutsche Nationalbibliothek
The Deutsche Nationalbibliothek lists this publication in the Deutsche Nationalbibliografie; detailed bibliographic data are available on the Internet at <http://dnb.d-nb.de>.

© 2013 Wiley-VCH Verlag & Co. KGaA, Boschstr. 12, 69469 Weinheim, Germany

Composition Laserwords Private Ltd., Chennai, India
Printing and Binding Markono Print Media Pte Ltd, Singapore
Cover Design   Adam Design, Weinheim

Printed in Singapore
Printed on acid-free paper

Print ISBN: 978-3-527-33028-7
ePDF ISBN: 978-3-527-64570-1
ePub ISBN: 978-3-527-64569-5
mobi ISBN: 978-3-527-64571-8
oBook ISBN: 978-3-527-64568-8

Contents

Acknowledgment *XI*
Preface *XIII*

1   **Classification of Explosives**   *1*
1.1   Initiation Sensitivity   *1*
1.2   Size   *1*
1.3   Usage   *2*
1.4   Physical Form   *2*

2   **Explosive Science**   *5*
2.1   Introduction   *5*
2.1.1   Low Explosives   *5*
2.1.2   High Explosives   *5*
2.2   Initiation and Detonation   *6*
2.2.1   Mechanism   *6*
2.3   Propagation and Detonation   *7*
2.3.1   Propagation   *7*
2.3.2   Detonation   *8*
2.3.2.1   Ideal/Nonideal Detonation/Critical Diameter/Ideal Diameter   *9*
2.3.2.2   Detonation Pressure and Velocity   *9*
2.4   Reaction Chemistry in Explosives   *11*
2.4.1   Heat of Reaction   *11*
2.4.2   Rules of Hierarchy   *12*
2.4.3   Calculation of Oxygen Balance and Fuel Values   *12*
          References   *13*

3   **Ammonium Nitrate Explosives**   *15*
3.1   Introduction   *15*
3.1.1   Chronology   *15*
3.2   Design of Commercial Explosives   *16*
3.2.1   Importance of Oxygen Balance   *16*
3.2.2   Physical, Performance, and Safety Requirements   *17*

3.3   Tests   *17*
3.3.1   Ballistic Mortar Test   *18*
3.3.2   Trauzl Lead Block Test   *19*
3.3.3   Velocity of Detonation (VOD)   *20*
3.3.4   Gap Test and Continuity of Detonation Test   *22*
3.3.5   Aquarium Test   *23*
3.3.6   Double Pipe Test   *23*
3.3.7   Cylinder Test (Crushing Strength)   *24*
3.3.8   Plate Dent Test   *24*
3.3.9   Underwater Test (UWT)   *24*
3.3.10   Crater Test   *26*
3.4   Assessment of Safety and Stability Characteristics   *27*
3.4.1   Impact Test   *27*
3.4.2   Torpedo Friction Test   *27*
3.4.3   Accelerated Hot Storage (ageing Test)   *27*
3.4.4   Cold Temperature Storage Test   *28*
3.4.5   Thermal Stability Tests Using DTA and TGA Procedures   *28*
3.5   Summary   *29*
            References   *29*

4   **Ammonium Nitrate and AN/FO**   *31*
4.1   Introduction and History   *31*
4.2   Physical and Chemical Properties of Ammonium Nitrate   *32*
4.2.1   Basic Data   *32*
4.2.2   Decomposition Chemistry of AN   *32*
4.2.3   Phase Transition in AN and its Importance in Explosives   *33*
4.3   Manufacture of Ammonium Nitrate   *35*
4.3.1   Prilled Ammonium Nitrate   *36*
4.4   Ammonium Nitrate Fuel Oil Explosives   *39*
4.4.1   Background   *39*
4.4.2   AN/FO Manufacture   *39*
4.4.2.1   Mixing Process and Equipment   *39*
4.4.2.2   Continuous Process   *40*
4.4.2.3   Bulk Delivery Systems   *40*
4.4.3   Properties of AN/FO   *41*
4.4.3.1   Physical   *41*
4.4.3.2   Oil Absorbency and Porosity/Bulk Density/Crushing Strength   *41*
4.4.3.3   Resistance to Effect of Temperature Cycling   *44*
4.4.4   Characteristics of ANFO   *45*
4.4.4.1   Density/Strength   *45*
4.4.4.2   Strength of the AN/FO Explosive   *46*
4.4.4.3   Energy Content of AN/FO   *46*
4.4.4.4   Velocity of Detonation and Effective Priming   *47*
4.4.4.5   Mechanism of Detonation Propagation in AN/FO   *49*
4.4.4.6   Influence of Fuel   *50*

4.4.4.7   Effect of Moisture/Wet Boreholes/Water-Resistant AN/FO   *50*
4.4.4.8   Water-Resistant AN/FO   *52*
4.4.4.9   Increasing the Energy of AN/FO and its Fume Characteristics   *52*
4.4.5   Safety Considerations in AN/FO   *55*
4.4.6   Summary – AN/FO Explosives   *56*
4.4.7   Quality Checks   *56*
           References   *58*
           Further Reading   *58*

5   Slurries and Water Gels   *59*
5.1   Development   *59*
5.2   Design   *59*
5.2.1   Large-Diameter Packaged Product (Water Gels)   *60*
5.2.2   List of Ingredients   *60*
5.2.3   Small-Diameter, Cap-Sensitive Water Gels   *60*
5.2.4   Bulk Delivery Product   *61*
5.2.5   Basic Concepts of Formulation   *61*
5.2.5.1   Oxygen Balance   *61*
5.2.5.2   Thumb Rules for Design   *62*
5.2.5.3   Role of Water   *63*
5.2.5.4   Basic Composition and Process   *65*
5.3   Process Technology   *66*
5.3.1   Batch Process   *66*
5.3.2   Continuous Process   *68*
5.3.3   Packaging Systems   *68*
5.4   Quality Checks   *71*
5.4.1   Raw Materials   *71*
5.4.2   End Product Specification   *73*
5.4.2.1   Development of New Formulations   *73*
5.4.3   Role of Aluminum in Water Gels and Slurry Explosives   *74*
5.4.3.1   Atomized and Flake Powders   *74*
5.4.3.2   Aluminum Water Reaction   *78*
5.4.3.3   Important Tests for AL Powder for Use in AN-Based Water Gel
           Explosives   *80*
5.4.4   In-Process and Finished Product Checks   *84*
5.4.4.1   Oxidizer Blend Composition   *84*
5.4.4.2   Solid Ingredients   *85*
5.4.4.3   Liquid Ingredients   *85*
5.4.4.4   Mixing   *85*
5.4.4.5   Packing   *86*
5.4.5   Performance Tests   *86*
5.4.6   Safety Tests   *87*
5.4.6.1   Gap Test/COD   *87*
5.4.6.2   COD   *87*
5.4.7   Storage Tests   *87*

5.4.8 Gel Condition Evaluation 89
5.4.9 Waterproofness Test 90
5.4.10 Effect of (Hydrostatic) Pressure 90
5.5 Process Hazards (Dust Explosions/Fire Hazards/Health Hazards) 91
5.5.1 Slippery Floor 92
5.5.2 Health Hazard 92
5.6 Role of GG 92
5.6.1 Application in Water Gels and Slurries 93
5.6.2 Specification of Typical GG Used in Water Gels 94
5.6.3 Cross-Linking 95
5.6.4 Mechanism of Hydration 96
5.7 Permissible Explosives 98
5.7.1 Design Criteria 98
5.7.2 Tests for Permissibility 99
5.7.3 Other Tests requirement 100
5.7.3.1 Deflagration Tests 100
5.7.4 Behavior of Water Gels in Permissibility Tests 101
5.7.5 Toxic Fumes and Typical Formulation 104
5.7.6 Strength of Permissible Water Gels 104
5.8 General Purpose Small-Diameter Explosives (GPSD) 105
5.8.1 Design Criteria and Composition 105
5.9 Sensitizers 106
5.9.1 Inorganic 106
5.9.2 Organic Sensitizers 107
5.9.3 Air/Gas/Synthetic Bubble Sensitizers 108
References 111
Further Reading 112

6 **Emulsion Explosives** 113
6.1 Introduction 113
6.2 Concept of Emulsion Explosives 113
6.3 General Composition of Emulsion Explosives 114
6.4 Structure and Rheology 115
6.5 Composition and Theory of Emulsion Explosives 117
6.6 Manufacture 118
6.6.1 Types of Emulsion Explosive Products 118
6.6.2 Manufacturing Process 118
6.6.2.1 Batch Process 119
6.6.2.2 Semicontinuous Operation 119
6.6.2.3 Fully Continuous Process 120
6.6.2.4 Critical Equipment for Production of Emulsion Explosives 121
6.6.2.5 Pumps 122
6.6.2.6 Packaging Equipment for Emulsion Explosives 122
6.6.3 Raw Material for Emulsion 123

6.6.3.1   Fuel Blends   *123*
6.6.4   Sensitizing in Emulsion Explosives   *125*
6.6.4.1   Air Entrapment or Occlusion while Emulsification by Mechanical
            Agitation   *125*
6.6.4.2   Chemical Gassing   *125*
6.6.4.3   Hollow Particles   *126*
6.6.5   Crystal Habit Modifiers   *127*
6.6.6   Emulsion Promoters   *128*
6.6.7   Emulsion Stabilizers   *128*
6.6.8   Emulsion Chemistry and Understanding Emulsifiers: Key to Good
            Emulsions   *129*
6.6.9   Concept of HLB and Its Use in Emulsification   *133*
6.6.9.1   Effect of Factors on Stability of Emulsions   *135*
6.6.10   Polymer Systems in Emulsion Explosives   *138*
6.7   Quality Checks   *139*
6.7.1   Raw Materials   *139*
6.7.2   Process Audit   *140*
6.7.3   Special Tests for Emulsions   *141*
6.8   Explosive Properties of Emulsion Matrix/Explosives   *142*
6.8.1   Channel Effect   *144*
6.9   Permissible Emulsions   *145*
6.10   General Purpose Small-Diameter (GPSD) Emulsion Explosives   *147*
6.11   Bulk Emulsions   *149*
6.12   Heavy AN/FO   *151*
6.13   Packaged Large-Diameter Emulsion Explosives   *153*
         References   *154*
         Further Reading   *155*

7   **Research and Development**   *157*
7.1   Areas of Interest   *158*
7.2   Development Work and Upscaling   *159*
7.3   Management of R&D   *161*

8   **Functional Safety during Manufacture of AN Explosives**   *163*
8.1   Introduction – Personal View Point on Safety   *163*
8.2   Safety Considerations in AN Explosives   *165*
8.2.1   In AN/FO   *165*
8.2.2   In Slurries and Emulsions   *166*
8.2.3   Electrostatic Ignition   *167*
8.2.4   Lightning Protection   *168*
8.2.5   Runaway Reactions   *168*
8.2.6   Venting as Means of Protection   *170*
8.2.7   Explosion Suppression Technology   *171*
8.3   Explosion Hazards in Equipment   *172*
8.3.1   Hazards Associated with Pumping of Explosives   *172*

8.3.2   Possible Hazards during Packing   *175*
8.4   Concluding Remarks   *176*
            References   *177*

9   **Economics of AN-Based Explosives**   *179*
9.1   In Manufacture   *179*
9.2   In Applications   *181*
9.2.1   Condition of Explosive   *182*
9.2.2   Coupling and Priming   *183*
9.2.3   Stemming and Confinement   *184*
9.2.4   Explosives–Rock Interaction   *185*
9.2.5   Explosives Energy Optimization in Blasting   *185*
9.3   Blast Design   *186*
9.4   Influence of Explosives in Underground Mining   *190*
            References   *193*

10   **Current Status and Concluding Remarks**   *195*

            Appendix A   *199*

            Appendix B: Guidelines for Investigation of an Accident   *203*
B.1   Introduction   *203*
B.2   Detailed Inspection   *204*
B.3   Interviewing and Questioning   *205*
B.4   Collection of Samples   *205*
B.5   Examination of Witnesses   *206*
B.6   Examination of Dead/Injured   *206*

            Index   *209*

Acknowledgment

It is not easy to write an acknowledgment for a book production as the number of people involved could be very many. The contents of my book are a mix of my own thinking and experience, but after stimulating discussions with various people connected with the global explosives industry and practical data collected over many years. But to my mind the motivation to stay and do research in this field was provided by the inspiring personality of Prof. T. Urbanskii with whom I came into contact while working in IDL industries, Hyderabad. The final impetus to write a book came through Dr. Martin Preuss of Wiley-VcH. I owe a lot to my family for their encouragement and support. Apart from these the various persons with whom I interacted in the industry at one time or the other inspired me to try and seek some answers to the phenomenon of explosives but special mention has to be made to the great working atmosphere provided by IDL Industries (now known as Gulf Oil India) where I spent a greater part of my career and gained hands-on knowledge in the field of explosives.

I am grateful to all the above for their role in motivating me to write a book primarily intended for the new entrants to the field of civil explosives as it stands today. Hopefully, I have succeeded in satisfying the readers of my book in whatever they expected from the contents.

I also wish to thank Dr. Martin Graf-Utzman and his staff for assisting me in finalizing this book.

Preface

Ammonium nitrate (AN) explosives came into prominence in the last three decades (40 years) for civil applications as it provided a greater margin of safety to the manufacturer and end user as compared to the then popular nitroglycerine (NG) explosives. Due to rapid industrialization over the last 30 years, there has been a surge in mining and infrastructure activities which in turn has triggered high demand for civil explosives at all types of remote and tough locations. Mining methodology also underwent a change to cater to these requirements and huge open cast mines which need for their efficient operations large volumes of explosives delivered at mine site are operating in great numbers. Such enhanced requirements could only be satisfied by AN-based civil explosives which can be manufactured in high tonnages with a good margin of safety and low capital investment. Thus in many countries, manufacture and use of NG explosives were reduced drastically or abandoned, and AN explosives were used instead. Rapid development of AN-based explosives for all types of applications including underground gassy coal mines took place between 1970 and 1990 to fill in the void left by NG explosives and a number of patents appeared during this time. Over the years, however, the manufacture in industrial scale has settled down to fairly common practices and raw materials.

Despite the importance of the AN-based civil explosives today, there has not been much written and published about these explosives in detail perhaps because of their unglamorous nature as compared to military explosives and propellants. It is my intention to fill this gap by devoting the contents of this book exclusively to the technology of manufacture of AN civil explosives. This book will deal with three such products – AN/fuel oil explosives (AN/FO), slurry and water gel explosives, and emulsions explosives, in great detail as they comprise nearly 90% of the explosives used worldwide in civil sector.

Much has been published about the chemistry and science of explosives as well as test methods employed to determine their characteristics. It is my intention not to repeat these but mention only the most important aspects applicable to AN explosive under consideration here. On the other hand, the book will concentrate on providing valid data and sensible manufacturing guidelines based on the author's hands-on experience of more than 35 years in this field. There is no attempt here to bring into print any kind of proprietary information and "tricks of the trade" being practiced in the industry. The author hopes that the contents will benefit the persons engaged in the industry to have a better understanding of the role of the critical factors involved in manufacturing good explosives in a safe way. It is also the fond hope of the author that through this book young and fresh minds will get stimulated to take up research in this subject, which has been woefully very meager, and contribute toward better understanding of the basics leading to safer and hopefully cheaper products in keeping with current environmental conditions.

I feel the chapters describing the critical role of raw materials like guar gums, aluminum powders, emulsifiers, processing and packing technology, and optimum utilization of explosives energy in field applications will be of great interest to the reader. Strangely while the science of explosives includes aspects of importance from complex subjects like thermodynamics, colloid chemistry, powder metallurgy, mixing technology, and detonation physics, the commercial manufacture of AN explosives for civil application has reduced to a fairly low technology, high volume industry where know-how is supreme and know-why is of low importance. Hence there is no theoretical and mathematical approach of the subject in the book but attempt is made to demystify and simplify concepts of explosive phenomena so as to enable those performing routine jobs in this industry to understand and appreciate more their occupation, thereby deriving more intellectual satisfaction.

The contents of the book will be of interest to persons engaged in the civil explosives industry in all its aspects such as manufacturing, quality assurance and safety, scientists in research establishments, statutory authorities in the field of civil explosives, individual consultants to the explosive industry, managers in the mining industry, and so on. The contents of the book could be used to write up a production and safety manual as also for troubleshooting in existing operations. The blasting engineer may also be able to use its chapter on application to derive the maximum benefit from the use of the explosives.

The book, after exploring the evolution of these three types of explosives, will contain individual chapters dealing with science and technology, manufacturing, safety, and future R&D work needed.

Individual chapters are exclusive to the type of the explosive being discussed. The three major explosives dealt with are AN/FO, slurry/watergels, and emulsion explosives.

Individual chapters describe the following:

- Classification/types/characteristics/definition
- Explosives science
- Raw materials and their role
- Formulation techniques and components
- Manufacturing technology
- Quality – concept and assurance
- Safety – understanding and practice

General topics of interest are contained in

- future R&D work,
- comparison with NG explosives,
- comparison between ANFO, slurries, and emulsions,
- economics of manufacture,
- applications,
- economics for the end user, and
- references and bibliography.

Erode G. Mahadevan

1
Classification of Explosives

Explosives are classified into different types and categories in various ways depending on their usage, sensitivity to initiation, and finished product packaging.

1.1
Initiation Sensitivity

- *Cap-sensitive explosives*: The explosive can be fully detonated with a measurable unconfined velocity in low diameters (1 inch) by initiating with a single detonator of No. 6 strength, which is the lowest strength detonator commercially made.
- *Booster-sensitive explosives (blasting agent)*: This type of explosive is detonated only when a booster of sufficient power is used to initiate it. These boosters are made of high explosives like pentaerythritol tetranitrate (PETN) and trinitro toluene (TNT) and are much more powerful than detonators (Figure 1.1).

Explosives are further classified into primary, secondary, and tertiary explosives depending on its level of sensitivity to external stimuli. Standardized testing evaluates the sensitivity in terms of friction, impact, heat, shock and based on these results, explosives are classified accordingly.

Nitroglycerine (NG) is very sensitive and classified as a primary explosive. TNT/RDX/dynamites are secondary explosives. These are relatively safe for handling and can be handled in large-scale production plants with acceptable degree of safety. Ammonium nitrate (AN) explosives are the least sensitive and come in the tertiary explosives group. Even though they may have higher detonation velocities and pressure than NG explosives, they are much safer to produce in very large quantities.

1.2
Size

- *Small diameter explosives*: These are usually explosives made in diameter of 7/8 to 2 in. and generally cap-sensitive.
- *Medium diameter explosives*: These are usually explosives made in diameters of 3–5 in. and are booster-sensitive only.

Ammonium Nitrate Explosives for Civil Applications: Slurries, Emulsions and Ammonium Nitrate Fuel Oils,
First Edition. E.G. Mahadevan.

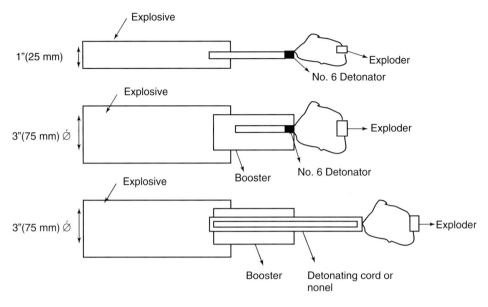

Figure 1.1 Types of initiation.

- *Large diameter explosives*: These are usually explosives made in diameter of 5–10 in. and are booster-sensitive only.

The boosters are in turn set off by either detonator or by a coil of detonating cord wound over and through it (see Figure 1.1).

1.3
Usage

The explosives are also classified into general purpose and permissible categories.

- *General-purpose explosives*: Usually in small diameter and cap-sensitive used for quarrying, tunneling, and cannot be used in underground coal mines.
- *Permissible explosives*: Cap-sensitive small-diameter explosives from 1 1/4 to 1 5/8 in. diameter, allowed by authorities for use in underground coal mines. Depending on the degree of gassiness (methane emission) found, there are further subclassifications. These differ from county to county depending on the test Procedures used.

1.4
Physical Form

Classification according to physical form of end product is as follows:

- *Cartridged explosive:* Here the explosive is in the form of cylindrical package, enclosed in paper or polythene tubings (flexible or rigid).
- *Pumpable explosives (bulk explosives):* Here the explosive is in the form of a flowy material and is capable of being pumped, augured, or poured. There is no outer packaging at all and the product is directly moved into the bore hole using bulk delivery trucks.

Any material which cannot be fully set off with a measurable velocity of detonation (VOD) either by detonator or by detonation is considered as "nonexplosive" in nature. However such nonexplosive material can be converted into an explosive by increasing its sensitivity.

2
Explosive Science

2.1
Introduction

Explosives are known in practice as substances which work on the surroundings when they are set off. In the open area, their effectiveness is much less than under confinement because most of the work is done by expanding gases. Gas-producing event can be due to burning (deflagration) or explosion and detonation. One usually differentiates by the reaction velocities and pressures achieved in each of the phenomena. Thus while in deflagration, reaction velocities are much slower than velocity of sound and the pressures attained are in the range of bars. In detonation, the reaction velocity, which produces gas due to chemical reaction of the explosive with its own ingredients or air, exceeds the speed of sound in the material itself; thus there is a supersonic shock wave produced. The wavefront travels in advance of the release of expanding gases. The shock energy has a high peak pressure but is transient, whereas the gas energy is a longer lasting event though lower in peak pressure attained (see Figure 2.1) [1].

2.1.1
Low Explosives

Deflagration and fast-burning substances which still perform some amount of work through release of gas are classified as low explosives. Black powder is a typical example. Reaction velocities are normally in the range of 600–1000 m/s (see Figure 2.2) [1].

2.1.2
High Explosives

Velocity of detonation (VOD) are in excess of 1800 m/s. Most commercial explosives and especially the ammonium nitrate (AN) based belong to the high explosives category due to their high detonation and gas pressures.

Explosives can also be classified as homogeneous and heterogeneous. Usually primary and secondary explosives are present in the former, whereas in the latter tertiary explosives which are mixtures of chemicals are found.

Ammonium Nitrate Explosives for Civil Applications: Slurries, Emulsions and Ammonium Nitrate Fuel Oils,
First Edition. E.G. Mahadevan.
© 2013 Wiley-VCH Verlag GmbH & Co. KGaA. Published 2013 by Wiley-VCH Verlag GmbH & Co. KGaA.

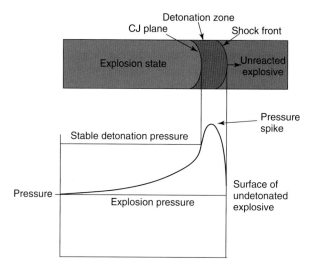

Figure 2.1 Schematic representations of zones and pressure variations along a detonating explosive charge.

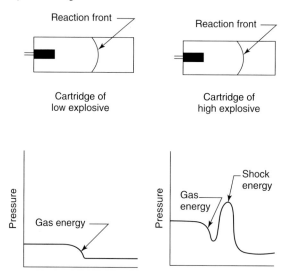

Figure 2.2 Pressure profiles for low and high explosives.

2.2
Initiation and Detonation

2.2.1
Mechanism

In order to perform, explosives need to be initiated. The extent of initiation needed depends on the sensitively of the explosive. Tertiary explosives – AN-based explosives, can be initiated by either a detonator or a booster like PENTOLITE. It is worthwhile to understand the basics of the phenomenon of initiation because of its importance, it being the first step toward detonation which is the ultimate goal.

In the case of most commercially used AN-based explosives, there is no self-explosive ingredient added on and the detonation is based on the rapid decomposition of AN and its reaction with the surrounding materials. As mentioned earlier if the chemical reaction is completed within the shock velocities, then detonation has occurred. Theoretical treatment of the mechanism of initiation abounds in literature [2–7] but while there are variations, there is also unanimity that hotspots within the explosive are essential for initiation and propagation of the detonation wave [8].

The hotspot theory has been to a great extent supported by experimental data obtained while formulating and testing AN explosives. The initiation is achieved by compression of the unreacted chemical (AN) causing local shear failure and inelastic flow that creates hotspots, which in turn sustain the chemical reaction by supplying intense heat [9, 10] in excess of the losses that could arise due to side effects as the wave travels. Several mechanisms postulated for creation of the hotspots and causing initiation are adiabatic compressions of voids through (i) shock pressure wave, (ii) friction, (iii) shear, and (iv) deformation.

These are achieved by shock/mechanical impact of hot metallic fragments in practice when a detonator shell explodes inside a tertiary explosive. On the other hand boosters initiate through pressure wave compression. Difference in initiation capability of No. 6/No. 8 for some explosives is due to higher brisance of No. 8 detonator containing higher quantity of explosive and its ability to accelerate metal fragments. This depends on the base charge, higher density, particle size, and physical status.

However, all the above mechanism finally lead to a thermal event (increase in temperature), which accelerates the decomposition reaction into a runaway situation. The latest theory finding acceptance is that inelastic flow of the explosive under the influence of detonation wave from the initiation is able to create a lower stress required for easier initiation.

2.3
Propagation and Detonation

2.3.1
Propagation

Propagation can be defined as the event where in an explosive the detonation process goes on till all the explosive material is consumed.

Propagation in a cylindrical charge of explosive is from the end of initiation to the other end. This is the most common event as most explosives are used in cylindrical shape of varying lengths and diameters. The word propagation itself means continuation and hence for an explosive to succeed initiation/detonation propagation is a must. The mechanism of propagation has also been the subject of study by many [11, 12].

Initiation *per se* does not ensure propagation. Attaining a steady-state detonation with a constant measurable velocity is a sure sign of propagation. For propagation to continue there must be continuity in the decomposition of the explosive till the end. This is achieved best when the losses due to conduction to the sides in a cylindrical charge is much less than the heat evolved as the wave travels through a cylindrical charge. There is net heat energy gain which accelerates the chemical reaction following Arrhenius equation. This heat generation is a self-sustaining multiplying situation as long as the heat transfer rate cannot keep pace with the exothermic heat produced by the reaction. Pressure and temperature rapidly increase leading to detonation in the direction of the shock wave. These conditions are especially critical in small diameter explosive charges.

2.3.2
Detonation

From initiation to propagation to detonation completes the chain of events as far as the explosive is concerned. Most of the postulations and theoretical calculations were made in the period of 1950–1970 and correlation with experimental data attempted. The study was mainly confined to pure explosive compounds. A certain degree of confidence and uniformity in prediction of behavior was established. However heterogeneous mixtures such as ANFO/slurries/emulsions were studied later since these were backbone of commercial blasting operations. Studies while not carried in depth were sufficient to establish guidelines to determine optimum behavior in practice. AN/FO being the simplest explosive consisting of two or three components was examined at great length and useful parameters were established.

Study of the detonation phenomenon has shown that the term *detonation* can be applied to the event where an explosive on receiving an impulse undergoes rapid and continuous exothermic chemical reaction. If the rate of reaction exceeds the velocity of sound in the reacting material, a supersonic shock wave is established. This shock wave propagates and ends when all the explosive material has reacted. The shock wave also produces a sharp high pressure which peaks and subsides as it propagates. The shock wave being a compression wave also produces heat in the body of the explosive. Just behind the shock wave is the reaction zone where the chemical decomposition is taking place (see Figure 2.3). Further due to this decomposition and combination with oxygen, gaseous products are evolved and they rapidly expand due to heat. This rapid expansion of the products of reaction from materials making up the explosive is the major work force employed for deriving a heaving effect while blasting. The energy in a shock wave as estimated, calculated, and measured is different. Bubble energy for different explosives and in absolute terms could be 15–20% for the shock energy and 80% for Total energy. The partition of total energy between shock and bubble is also different for different explosives (see Table 2.2).

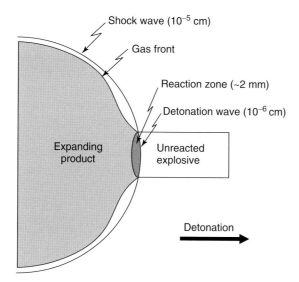

Figure 2.3 Structure of Detonation front [11].

2.3.2.1 Ideal/Nonideal Detonation/Critical Diameter/Ideal Diameter

The event of detonation and measurements thereof have shown that VOD is steady throughout and for the same explosive does not increase or decrease unless the diameter increases or decreases within limits. Where there is no further increase in VOD, however much the cylindrical cross section of the explosive increases; then it is termed as a *state of ideal detonation*. Conversely where the VOD is affected by the diameter of the explosive, nonideal detonation exists. The difference in diameter for these two states for most commercial explosives can be significant and has to be kept in mind during blasting operations.

Detonation has also thrown up the fact that behavior of explosive changes subject to its diameter. The smallest diameter of a cylindrical charge at which steady-state detonation (unconfined) takes place consistently is termed as the critical diameter (CD). Below CD the explosives behavior is unpredictable both in terms of initiation and performance. The lowest diameter at which ideal detonation is achieved (when max VOD is attained) is termed as IDEAL diameter (ID). It would make sense if the blasting is carried out at this diameter but may not always be possible due to other constraints.

The values of CD and ID are reported for explosive charges using standardized initiation and the charges are not confined. The values are higher for confined explosives but can again vary depending on the degree of confinement. This also can be established while experimenting to obtain comparable results (Table 2.1).

2.3.2.2 Detonation Pressure and Velocity

Detonation pressure and detonation velocity are two of the most important parameters for an explosive used extensively to judge its closeness to ideal detonation.

Table 2.1 Velocity of detonation of various explosives.

Velocity of detonation (in m/s)				
	Water gel 25 m diameter	Blasting gelatin (80%) 25 mm diameter	AN/FO 100 mm diameter	Emulsion 25 mm
Unconfined	3500–4000	4000–4200	2200–2500	4500–5000
Confined	5000–5300	5000–5300	4000–4200	5200–5500

The two terms are interlinked in the form of a cause and effect situation and this relationship is given in Eq. (2.1).

$$(P_D^2)/4 \text{ where } P_D^2 \alpha D^2$$
$$P_D = P_D^2/R + 1 \tag{2.1}$$

where P is density of the unreacted explosive and 4 is a calculated value. R is the ratio of specific heats of the detonation products given (i.e., 3 in most cases). There are also theoretical calculations for estimating the detonation velocity. Readers interested in studying these and also to understand the detonics of explosion can look into relevant chapters of the book by Cooper [12].

For the purposes of this book, the commercial explosives' velocity detonation is a measured value which can be easily worked out by experimental set ups ranging from the classical D'Autriche method to electronic measurements. The latter are very useful in tracing the VOD as the detonation wave progresses and are often used for measurement in a borehole during blasting.

While it is true that volumes have been spoken and written about detonation, pressure/VOD, in practice it is best used as a figure to indicate the state of the explosive in comparison to (i) ideal or theoretical maximum and (ii) figure at the start of a life cycle of the explosive.

It is also to be kept in mind that explosive performance depends not only on the detonation pressure but also on other work forces such as those generated by expanding gases formed during decomposition of the explosive ingredient(s). Several studies including underwater energy measurements estimate that out of the total energy expended by an explosive only 15–20% is partitioned as shock energy (from detonation pressure) and balance from gas/bubble energy from expanding gases.

However, there is no doubt that the shattering effect (or ability to impart movement to metal objectives is totally related to detonation pressure. Since PD is connected directly to VOD (D) it is obvious that factors affecting VOD such as density, diameter, particle size, and homogeneity of the explosive will also influence detonation pressure.

In general the effect of density has been theoretically and practically studied in pure explosives and mixtures. It has been established that PD is proportional to

Table 2.2 Partition of energy.

Explosive	Density	Bubble energy	Shock energy	Energy split
AN/FO	0.80	2.44	0.77	76/24
Emulsion	1.3	1.68	0.52	75/25
Water gel	1.2	2.2	0.5	81/19
	1.36	2.03	0.57	78/22

P^2 and hence D is proportional to P. Increase of density increases the detonation velocity and consequently the detonation pressure and vice versa. The fact that an increase in bulk density beyond a certain point also brings in reduced sensitivity to initiation limits the usefulness of the above findings in practice. Behavior is contradictory for pure explosives, and those containing AN as the main component. The influence of particle size/homogeneity while understood better in pure explosives has not been so well established in mixtures based on AN. Decreasing particle size does not necessarily lead to higher sensitivity. Table 2.2 provides VOD and energies of some well-known types of commercial explosives [13].

2.4
Reaction Chemistry in Explosives

2.4.1
Heat of Reaction

While we have previously discussed the influence of pressure developed in a detonation, we take a look now at the chemical reaction in the explosive. It is recognized that an explosive undergoing detonation releases a lot of heat (energy). Many of the injuries suffered in accidents involving explosives are burn injuries suffered due to intense heat. In practice, this huge release of heat energy goes in accelerating the chemical reaction and enabling pressure build up (spike) in a super fast time scale.

The heat of reaction also in certain cases termed as *heat of detonation* is the energy difference between the reactants and products. Enthalpy is the measure employed to quantify the energy in a chemical substance (Eq. (2.2)).

$$\Delta H(R) = H(P) - HR \tag{2.2}$$

where $\Delta H(R)$ = heat of reaction,
$H(P)$ = enthalpy of products, and
HR = enthalpy of reactants.

Although the heat of reactions can be empirically found [14], ΔH values are somewhat inconsistent as the type and quantity of products obtained depend on several parameters connected with the explosive itself and conditions under which it detonates such as initial density, degree of confinement, composition including the presence of metals, particle size, and structure. For experimental determination of product gases, qualitative and quantitative tests are preferred, but under field conditions of blasting they are not at all easy to gather.

2.4.2
Rules of Hierarchy

A reasonable approximation of ΔH is obtained by following certain rules of hierarchy consistently so that products obtained are close to ideal. The rules state that

1) All nitrogen becomes N_2.
2) All available oxygen goes first to convert hydrogen to water.
3) Leftover oxygen from step (2) converts carbon to CO.
4) Leftover oxygen from step (3) converts carbon to CO_2.
5) Leftover oxygen from step (4) is present as O_2, available for use in secondary reactions.
6) Any leftover carbon becomes solid residue.

Some researchers advocate conversion of all remaining oxygen after formation of water to go toward formation of carbon dioxide. This method yields higher estimate of heat of reaction than the first one. ΔH values are well published by many sources [5, 11, 12]. Riggs has suggested a novel method of estimating the reaction products to a value near to that obtained by more sophisticated methods. Here the first use of oxygen is to form CO, and the remaining O_2 is split between CO and H_2 to produce CO_2 and H_2O. Since enthalpy values depend on product concentrations, it is assumed that calculation of heat energy released is an approximation. The detonation products emerging from the reaction zone of a detonation wave have transient existence and react with other components and form new products.

2.4.3
Calculation of Oxygen Balance and Fuel Values

The concept arises out of the oxidation reaction, which is basis for explosive energy. The explosive energy tied up with production of heat/gaseous products depends on the oxygen balance present in the explosive. This is applicable to chemical mixtures acting as an explosive composition also.

Knowledge of oxygen balances influence on explosive properties will help in formulating an explosive with optimum performance.

Heat of reaction of an explosive reaches its maximum when it has just enough oxygen to convert all its fuel (C, H) to this higher oxidation state (CO_2, H_2).

Calculations of oxygen balance are well standardized as shown in Eq. (2.3). Using Eq. (2.3), the oxygen balance for ammonium nitrate is derived to be 20.

$$\%OB = \frac{O - 2C - \frac{1}{2}H}{MW} \times 1600 \tag{2.3}$$

where

O = oxygen,

C = carbon,

H = hydrogen, and

MW = molecular weight.

We have seen that for maintaining oxygen balance of an explosive we may have to add fuel if it is oxygen-rich or add oxygen-giving substance (oxidant – oxidizer) if it is oxygen-deficient. In order to calculate fuel values, the same hierarchical process is used in reverse. The molecular weight of compound is divided by the oxygen deficiency using Eq. (2.4).

$$\text{Fuel value} = \frac{MW}{2C + \frac{1}{2H} - O} \tag{2.4}$$

where

MW = molecular weight,

C = carbon, and

H = hydrogen.

Oxidizer value is calculated using Eq. (2.5).

$$\text{Oxidizer value} = \frac{MW}{O - 2C - \frac{1}{2H}} \tag{2.5}$$

where

MW = molecular weight,

C = carbon, and

H = hydrogen.

It is to be noted that these values are not always correct as derived from the above formulas as other competing reactions like hydrogen combining with chlorine or sodium getting oxidized and giving a solid residue are not taken into account.

References

1. Konya, C.J. (1995) *Blast Design*, Intercontinental Development Corporation, Montville, OH.
2. Fordham, S. (1966) *High Explosives and Propellants*, Pergamon Press, Oxford.
3. Urbanski, T. (1964–1984) *Chemistry and Technology of Explosives*, vol. I–IV, Pergamon Press, Oxford.
4. Cook, M.A. (1958) *The Science of High Explosives*, Rheinhold Publishing Corp, New York.
5. Agrawal, J.P. (2010) *High Energy Materials*, Wiley-VCH Verlag GmbH, Weinheim.
6. Cook, M.A. (1974) *The Science of Industrial Explosives*, IRECO Chemicals, Salt Lake City, UT.
7. Johansson, C.H. and Persson, P.A. (1970) *Detonics of High Explosives*, Academic Press, London.
8. Bowden, F.P. and Yoffe, A.D. (1952) *Initiation and Growth of Explosives in Liquids and Solids*, Cambridge University Press, Cambridge.
9. Taylor, J. (1952) *Detonation in Condensed Explosives*, University Press, New York.
10. Zeldovich, J.B. and Kompaneets, A.S. (1960) *Theory of Detonation*, Academic Press, New York.

11. Riggs, R.S. *Elements of Explosive Behavior*, Jet Research Center, Arlington.

12. Cooper, P.W. (1996) *Explosives Engineering*, Wiley-VCH Verlag GmbH, Weinheim.

13. Cameron, A.R. and Torrance, A.C. (1990) Underwater evaluation of the performance of bulk commercial explosives. SEE Annual Conference, Orlando, FL.

14. Kubota, N. (2002) *Thermo Chemical Aspects of Combustion*, Wiley-VCH Verlag GmbH, Weinheim.

3
Ammonium Nitrate Explosives

In this book, we deal with ammonium nitrate/fuel oil (AN/FO), slurries, and water gels, emulsions. These types of explosives are containing AN in one form or the other and in general when commercially manufactured do not contain self-explosive ingredients. It is possible that while formulating small-diameter cap-sensitive products, borderline compounds like methyl amine nitrate could be included for obtaining higher sensitivity. Bulk or large-diameter products do not include even such compounds and hence are classified as *blasting agents* with even less stringent rules of storage and transportation.

3.1
Introduction

3.1.1
Chronology

It is claimed that even 150 years ago, AN was known as a *compound* which could be used in explosives or possessed explosive properties. However nitroglycerine (NG) explosives dominated this scene till about 1950. Even during the golden period of NG explosives, AN was used as an ingredient in so-called special gelatins to provide for the reduction in the NG content and to reduce sensitivity to manufacture and handling. AN used was in crystalline form. The first commercial application of AN as major ingredient in explosive happened in 1950 when AN was mixed with carbon as fuel and used in large boreholes. Subsequently in USA the fertilizer industry expanded rapidly and AN in the form of prills became available. Initially the same type of prills were used both by the fertilizer and explosive industry, but subsequently due to the difficulty in initiation and propagation even in large boreholes search into identifying properties leading to more sensitive prills began in earnest and soon a set of specifications were developed for prills useful to the explosives industry. These are given in detail in the later part of the book but in general it was for a lower density (LD), porous structure and for thermal and physical stability, flowability. Ammonium nitrate mixed with fuel oil became a dominant product and use of NG explosives became reduced and practically no NG-based explosives was used in large-diameter blasting in USA by mid-1960s and

Ammonium Nitrate Explosives for Civil Applications: Slurries, Emulsions and Ammonium Nitrate Fuel Oils, First Edition. E.G. Mahadevan.
© 2013 Wiley-VCH Verlag GmbH & Co. KGaA. Published 2013 by Wiley-VCH Verlag GmbH & Co. KGaA.

this spread to other parts of the world subject to availability of prills. The inefficiency of AN/FO in smaller diameters and in watery holes gave rise to evolution of slurry explosives and later emulsions in a big way.

The research group under Prof. M. A. Cook in Utah, USA, did pioneering work [1] in putting slurry explosives in the forefront and subsequently water gels came on the scene for packaged and cap-sensitive explosives for all types of applications including coal mines. The difference in slurries and water gels was basically in the physical structure of the explosives. While slurry was in a fluid form, water gels were semisolid once packed. The ingredients were AN as the major component present as a supersaturated aqueous solution. Special ingredients were added to get the desired physical consistency and were sensitized in different ways [2]. While AN/FO explosives consisted of only two to three ingredients, slurries/water gels had 8–10. Certain deficiencies noticed in water gels such as their diminishing sensitivity to initiation at cold temperatures, greater tendency toward nonideal detonation, irregular behavior in watery holes with time necessitated further search and emulsion explosives came into being in many places like USA, Sweden, India, South Africa, and China. The emulsions started with the concept of AN/FO in that they used AN and fuel oil but more intimate contact between them was built up through emulsification in water-based dispersions. Emulsion explosives have become extremely sought after as they are very amenable to high-speed pumping so that boreholes could be loaded *in situ* very fast. Also the velocity of detonation (VOD) in small diameter and critical diameter measured in well-made emulsion explosives indicate closeness to ideal detonation not achieved in other types of AN explosives indicating a more homogeneous structure and intimate contact between particles of oxidizer (AN) and fuel through the emulsification process.

Thus currently the scenario in the commercial civil explosives is that for specialized small-diameter cap-sensitive explosives requirements, tailor-made packaged emulsion or water gel product is used. On the other hand for most large diameter blasting where large volumes are required, pumpable emulsions are preferred. Pumpable slurries are also still being used in a few locations depending on the closeness of the base plant.

3.2
Design of Commercial Explosives

3.2.1
Importance of Oxygen Balance

The basics of maintaining close to zero oxygen balance (OB) is the first principle followed while designing good commercial explosives, whatever may be the type. This is to ensure that the explosive performs at its maximum potential and also that postdetonation fumes are least toxic.

In simple terms, the explosive consists of a mixture of oxidizer, fuel, sensitizer, and filler.

OB is calculated for any given formula by summing up of individual OB values obtained after giving due consideration to their individual quantities present in the total composition as a percentage.

The easiest way to theoretically calculate it is the case of AN/FO. AN has OB of +0.20 and FO has a value of −3.3 (both calculated using method discussed later). Balanced fully, the ratio of AN to FO would be 94.3/5.7. In the case of a typical water gel, it could be based on the formula

AN 62.5% (+20)	+12.5
Other oxidizers 8.0% (+45)	+3.6
Total oxygen available	+16.1
Contribution needed from fuel	−16.1

In case of packaged product, there has been much debate whether the wrapper, usually polyethylene, is taking part in the reaction or not. It becomes significant only in small diameters where up to 4–5% of explosive weight could be plastic. In case of large-diameter packages, the amount of plastic would be much less −0.5%. Formulators are well advised to keep the OB near zero without considering influence of packing material on the oxygen balance.

3.2.2
Physical, Performance, and Safety Requirements

The design of the explosive apart from the composition has to take care the physical and performance requirements and safety within the overall cost structure allowed by market consideration.

The physical requirements are guided by the type of explosive needed (cap-sensitive or booster-sensitive), its size (diameter and weight), its packaging, and delivery mode.

In case of AN/FO, the options for design are limited as the explosive is present only as a dry mix.

The safety considerations need to meet local safety rules for storage, transport, and manufacturer's own house standards and accordingly the design of the explosive is adjusted.

Details of the design criteria for the three individual products (AN/FO, slurries, and water gels, emulsions) are given separately in chapters devoted exclusively to them.

3.3
Tests

Ever since explosives came into existence, testing them has followed. Considerable time and application of mind to device tests to assess performance and safety properties of explosives have taken place and a number of such tests are known.

The tests in the earlier years [3] were oriented toward NG explosives as they were predominant during those years (up to 1960), but subsequently it was realized that the same tests could not be applied to non-NG explosives and new tests were established to encompass the newly developed AN explosives [4]. More and more importance was given to field performance of the explosives rather than their absolute properties and hence tests were designed to obtain correlation between the explosive characteristics and field performance.

A fairly comprehensive list of tests [5, 6] known and practiced are given in Table 3.1. Tests relating to strength and performance are described below:

While discussing about the strength of an explosive, the value will have to be qualified whether it is on weight or volume basis. While two explosives can be similar in weight strength, they can differ in their volume strength because of the difference in their density. For field applications, it is the volume strength (bulk strength) that is the deciding and critical factor.

3.3.1
Ballistic Mortar Test

This is one of the oldest well-known tests developed a century ago and used mostly for NG-based explosives. Blasting gelatin is used as a standard [3]. A typical setup of a ballistic mortar is shown in Figure 3.1.

A 10-g charge of explosive is detonated inside a mortar which is suspended from a well-anchored structure and is able to swing freely. When the charge explodes, the recoil measured in terms of maximum deflection is noted and compared with standard deflection obtained by use of a 10-g charge of blasting gelatin. Very reproducible results were obtained for NG-based explosives and the strength of these explosives was very often quoted as percentage of blasting gelatine (BG). However the test did not give reproducibility with non-NG explosives especially with AN/FO where in many instances the explosive did not fully detonate and traces were left behind. Even cap-sensitive emulsions and water gels showed variable results though reproducibility was better than that with AN/FO.

Factors affecting the test results and leading to variability were loading density, sensitivity, speed of the decomposition reaction, and stemming variability due to personal factor.

Table 3.1 List of tests.

Ballistic mortar test	VOD, COD, and air gap tests
Trauzl lead block	Plate dent test
Aquarium	Thermal stability
Double pipe test	Chemical compatibility
Cylinder test	Friction (torpedo)
Under water test	Freezing and thawing
Crater test	Cold temperature test
	Hot storage test

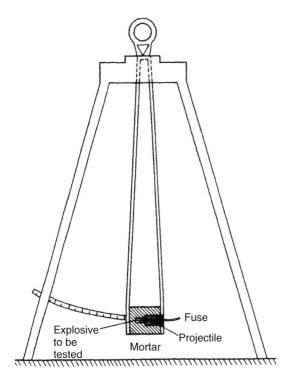

Figure 3.1 Ballistic mortar test. (Adapted from Mason and Aiken [4].)

Thus currently this test is not practiced on a routine basis with AN-based explosives and blasting agents. Several studies to correlate weight strength values obtained by this method with bubble energy (gas energy) and shock energy did not yield any meaningful relationship.

3.3.2
Trauzl Lead Block Test

This is another classical test developed in the early twentieth century and used for NG explosives and other molecular explosives. The test was used to measure and compare strengths while developing different formulations. Later it was also used as a type test to keep track of the product being manufactured and in storage.

Herein again it was found that the method did not yield reproducible results for non-cap-sensitive products, but for cap-sensitive explosives such as aluminized water gels the results obtained were good [7] and the values could be correlated with calculated energy and total underwater energy.

The test is simple and consists of detonating with a number 6 detonator, ten grams of explosive in a cast lead block with a 25-mm-diameter hole drilled in the center to a depth half way from the bottom (200 mm). The charge placed at the

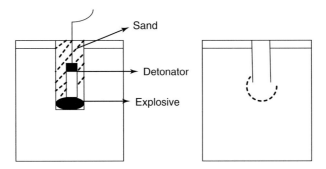

Figure 3.2 Lead block expansion test (Trauzel test).

bottom of the drilled hole is surrounded on three sides by lead block walls of sufficient thickness to prevent a blow hole. The remaining side (top) is stemmed with loose sand. For standardization and reproducibility, lead blocks cast from the same mold, composition, and dimensions are used. Also same type, quality, and quantity of fine sand for stemming and identical number 6 ordinary detonators are used. The explosive is contained in an aluminum foil of the same dimension every time. Once the charge explodes, the gases released expand the lead block and enlarge the volume inside (Figure 3.2).

The increase in volume of the hole in the lead block before and after detonation measured accurately by filling up to the brim with H_2O from a burette will give the volume increase only due to the explosive and is a measure of its weight strength. To remove the influence of the strength of the detonator used in the test, the volume of expansion in a lead block of a standard detonator is measured and used as a correction factor when obtaining the work done by the explosive.

Here again either BG or a standard NG explosive such as a special gelatin 80% is used for comparison.

While the ballistic mortar and lead block tests measure the work done by expanding gases behind the detonation wave, there are a few tests that measure the shock energy due to detonation wave as well as reaction velocity itself.

3.3.3
Velocity of Detonation (VOD)

It is a very common quoted figure to indicate the speed at which the reaction moves once the explosive is initiated. It also gives the condition of the explosive when its VOD is measured and compared with the maximum VOD attained or the ideal VOD achievable by that explosive. If the measured VOD is much lower than the latter, it means that the explosive is detonating at a reduced performance level and will not give the desired results in the borehole although confinement generally improves the VOD of most explosives by 10–15%. The diameter also has an influence on the VOD, and greater the diameter more is the VOD till a constant. VOD value is reached for that particular explosive (VOD_{max}). The VOD measurement as standardized is conducted in the open without confinement and

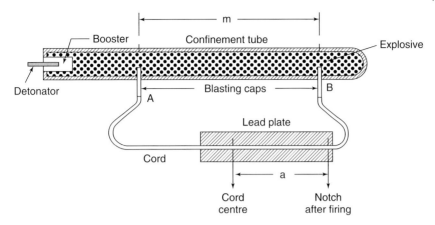

Figure 3.3 D'Autriche method for VOD determination.

in the diameter for which VOD is required to be measured. The initiation is also standardized. For cap-sensitive explosives, the number 6 detonator is used and for LD non-cap-sensitive explosives pentaerythritol tetranitrate/trinitrotoluene (PETN/TNT) booster is used, the latter being initiated either by a detonator or by a detonating fuse. The diameter of the booster should be as close as possible to the diameter of the cartridge being tested and the weight used is about 10% of the explosive weight.

The actual most used method for experimental determination of the VOD known as *D'Autriche method* consists of the setup as shown in Figure 3.3.

As the reaction proceeds after initiation, it sets off the two arms (probes) of the detonating fuse embedded at a fixed distance in the cartridge. The detonating fuse after getting initiated has two waves traveling in the opposite direction. They meet at a point and the collision is marked on a lead or aluminum plate. The distance of the mark on the plate from the midpoint of the DF used is a measure of the VOD of the explosive. A standardized value for the DF is used after it is calibrated and used in the equation:

$$D(X) = D \times \frac{m}{2a}$$

where:

$D(X)$ = the VOD of the explosive under test in m/s,

D = VOD of the calibrated detonating fuse used in the test in meters per second,

a = distance between the mark on the witness plate and center of the detonating fuse in centimeter, and

m = length between the probes in cm.

Over the years, the DF probe has been replaced by electronic devices based on fiber optics so the continuous measurement of VOD over the length of the cartridge can be measured. Electronic methods when they function well give very accurate results, but most routine testing is done by the classical method. However if the

VOD needs to be measured in a borehole, electronic methods are the only ones which can be used.

Some precautions needed to be observed while measuring VOD by D'Autriche method in that care should be taken in

1) placing the DF probes in the explosives exactly parallel so that it equals the distance measured between the punch holes made to insert the DF,
2) the distance of the mark on the plate from the initiation end is accurately measured, and
3) the total length the DF laid on or over the plate also to be accurately measured.

It also needs to be understood that while VOD is a good measure of the condition of the explosive, it is much less useful in predicting the behavior of the explosive in the borehole.

There is also the question of handling lead plates. To avoid this, the author has used Al strips from the Al rolls used for detonator manufacture (or scrap). The Al plate being much thinner can get cleaved if the DF is laid on it. Hence the DF is placed 1 in. above and the mark obtained is very clear.

3.3.4
Gap Test and Continuity of Detonation Test

This is a particularly useful test for propagation in small-diameter cap-sensitive explosives and to compare the sensitivity of different compositions. The gap test is well known and was used for NG explosives. The gap test showed for AN explosives lower values than NG explosives, and here again confinement increased the gap sensitivity significantly. In general metalized formulas showed higher gap sensitivity. The setup is to take two cartridges of SD (1 in. diameter) and separate them by a fixed distance (gap) in a paper rolled in cylindrical form (Figure 3.4).

The explosive is initiated from one end (donor cartridge). The condition of the receptor is observed after the detonation. The explosive is considered to have passed and jumped the gap if the receptor cartridge has completely detonated. In case there is doubt, the experiment is repeated. For greater accuracy, Bruceton up and down method is used [8]. In general confinement improves the gap sensitivity. When testing under confinement is done, thick steel tube which is not shattered by the test quantity of explosive is used. The gap test does not have much significance for booster-sensitive products and blasting agents since in a borehole the explosive is supposed to be in close contact.

The continuity of detonation (COD) test is also a very useful test to judge the ability of the explosive to propagate. Usually at least 1 m length of explosive column (cartridge placed touching each other) is initiated from one end and the explosive is considered to have passed the test if the entire column has detonated. A refinement of the test is to have a setup to measure the VOD in the last part of the column. If the VOD recorded here is the same as the usual VOD of the explosive, then the explosive has detonated at steady state. Here also confinement is found to improve the COD.

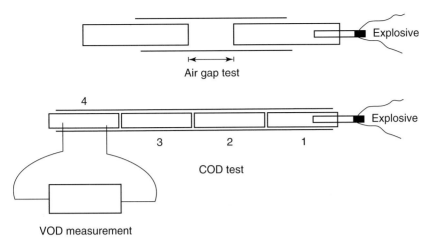

Figure 3.4 Test setup for VOD and air gap measurements.

COD tests with VOD measurement in the last part of the column can give an idea of the boostering required to maintain steady-state detonation.

COD values also give us information on the condition of the explosive at different periods of storage.

The precaution needed to be taken while performing gap/COD tests is to make sure that ends of the cartridge placed in contact have plane surface. Sausage-type cartridges with clipped ends should be cut to ensure symmetrical diameters since clipped ends reduce the diameter of the cartridge and can lead to erroneous results.

3.3.5
Aquarium Test

Here the explosive is detonated in small quantities in a glass tank filled with water, and movement of the shock wave in the water as it radiates is measured by a streak camera [9]. This test was found more useful for testing strength of detonators and military explosives as quantities tested need to be small to be accommodated in a reasonable size tank and at these quantities (a few grams) the AN explosives do not detonate ideally.

3.3.6
Double Pipe Test

It was developed with a view to correlate results in the test with field performance [10]. The deformation in the witness pipe in an arrangement consisting of two pipes of same diameter fitted one above the other, the above one filled with explosives, and detonated from above was thought to be a measure for the energy released by the test explosive. Again the results are a direct measure of the shock energy developed at peak pressure, but the heave energy from expanding gases is not reflected in the deformation of the witness pipe. Hence the performance of the explosive measured by this method does not fully indicate blast performance of the explosive in the borehole.

3.3.7
Cylinder Test (Crushing Strength)

A copper cylinder 50 mm in length and 25 mm in diameter is placed between two mild steel plates of 25 mm thickness. The top plate has a rim within which the explosive can be mounted vertically. The cartridge is also supported from outside. The initiation is from top. The detonation travels downward and presses the mild steel plate, which in turn crushes the copper cylinder. Extent of compression of the copper cylinder is a measure of the detonation pressure exerted on it by the explosive. The detonation pressure as we have seen earlier is dependent on the velocity of detonation. This test gives fairly reproducible results and can be used if correlation between VOD and detonation pressure needs to be obtained [5].

3.3.8
Plate Dent Test

This test is a modification of the cylinder test and is a measure of brisance or shattering strength of the explosive, which is again proportional to detonation pressure. All factors influencing detonation pressure else affect the brisance.

Here the explosive of given weight in a standard cartridge form is kept vertically in contact with a metal witness plate. The depression or crater left on the witness plate is a measure of brisance and the detonation pressure of the explosive. The method is best used for routine checking and comparison between explosives.

Results are reproducible provided the setup is same for every shot, especially the area of contact between the explosive and witness plate. This is easy to achieve for soft materials like slurries/water gel/emulsions. Any metal clips at the end should be removed and a rubber band should be used to close the end cartridge. Metal clips can fly for long distances and act as injurious missiles. The explosive cartridge length must be of sufficient length (greater than four charge diameters) to attain steady-state detonation before reaching the witness plate. Overboostering/overdrive needs to be avoided. The test was used initially for large-diameter explosives. In a method developed by the author (1973), a direct correlation was seen between dent depth and VOD.

3.3.9
Underwater Test (UWT)

Although this test requires elaborate setup and geographical support, it is considered as the one which comes closest to measure the total energy released by an explosive on detonation. The test was used earlier for evaluation of underwater depth charges, but later it was adopted for civil explosives evaluation. This method has been described and used by a number of researchers [11–14].

Figure 3.5 shows an arrangement usually used. In principle, an explosive charge of sufficient diameter and length to ensure adequately stable detonation in cylindrical form is detonated underwater at a depth and position away from the sides and bottom of the water body to ensure no interference from the solid terrain. Pressure transducers are mounted at a suitable distance/position from the explosive charge to record pressures from shock waves and bubble oscillations obtained after detonating the explosive.

Precautions to be taken to ensure reproducible results are as follows:

1) Adequate initiation of explosive taking into account its increased density due to hydrostatic pressure at the depth where it will be initiated.
2) Same geometry for all shots.
3) Pressure gauges to be calibrated often.
4) Enough space for bubble to develop fully.
5) If possible all shots to be fired at same explosive temperature.
6) Same degree of confinement for all shots.
7) Same initiation strength for all cap-sensitive explosives.
8) Same booster strength for all non-cap-sensitive explosives.

It has been found that the results give good reproducibility for both shock and bubble energy and correlation with calculated total energy of the explosive. However certain other field studies indicated that underwater behavior of explosives does not reflect fully its performance in field. The reason could very well be that while measurements are made in water, a homogeneous body with uniform properties, the rock is heterogeneous and its properties vary due to geological nature of the

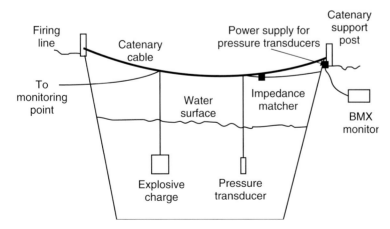

Figure 3.5 Arrangement of underwater test.

strata. Personally I would use the under water test for obtaining explosives' energy of new formulations (total energy and its components) and to get a good idea of where the newly formulated explosive stands in relation to a known product. The test can give information on the sensitivity of new formulations when fired under pressure developed underwater and in boreholes of great depth where the whole weight of a considerable length of column charge would be acting on the explosive at the lowest level, thereby raising its density leading to a possible reduction in its explosive properties. Because of the difficulty in setting up the test, it would not be possible to use it for evaluating routine production as a quality control tool.

3.3.10
Crater Test

It has been used for evaluating performance of explosives since many years and considered as closest to performance in actual field conditions. The procedure involves firing several single holes in a given rock type and a constant weight of explosive buried at varying depths and the crater volume measured. The latter will involve excavation of the crater and measuring its dimensions. It was found that at a particular depth for a certain explosive, no surface disturbance was seen and this depth is called *critical scaled depth*. Depth at which maximum crater volume is found is termed *optimum scale depth*.

A plot of scaled crater volume V/W (where V = crater volume, W = charge weight) versus scaled depth (d/w 1/3, where d = depth of burial of explosive) is used to obtain the critical scaled depth (minimum depth producing no surface disturbance), the optimum scaled depth (depth at which maximum crater volume is produced), and crater yield (V_{max}/W where V_{max} = maximum crater volume). A comparison of these values for a given explosive in a specific rock strata is a measure of its performance, and data obtained for different explosives can be compared for obtaining relative strengths and performance.

Slykhouse [15] found this to be a good method for comparison of different explosives, but there was no correlation with VOD or underwater energy. The rock strata and structure had great influence on the crater volume; nevertheless if the test is conducted using similar blast parameters as in the field, then one can get a good idea of the explosive's performance in practice.

Certain disadvantages are as follows:

1) No free face and hence crater volume does not give credibility to shock energy.
2) Excavation needed after every shot.
3) Difficult to find virgin test area.
4) Extrapolation of result from one type of rock to another can be misleading.

It may be more realistic to develop this test for evaluating cap-sensitive explosives where the entire setup will be on a reduced scale and will be easier to manage logistically.

3.4
Assessment of Safety and Stability Characteristics

3.4.1
Impact Test

As the name implies, this test measures the effect of impact according to a standardized setup and procedure. In principle, the impact test involves the study of behavior of the explosive material when a hammer of standard weight is made to fall from different heights onto the explosive kept on a hard metallic surface and observing whether the explosive initiated or not.

Most AN-based commercial explosives showed no reaction in this test except for compositions containing self-explosive ingredient.

3.4.2
Torpedo Friction Test

This test measures the friction sensitivity of explosives. A torpedo-like missile slides down and hits a glancing blow to the explosive placed on a hard base plate. The explosive is caught between the torpedo and the base plate and, unlike the fall hammer, the torpedo drags the explosive across the plate subjecting it to friction. A spark, or a small explosion or smoke when impacted by the torpedo, is a sign of positive action. The angle at which the torpedo strikes can be varied so that explosive is subjected to different frictional forces. An up and down evaluation of reaction to the test can be done if sufficient number of measurements are made, and this value is characteristic of that particular explosive and can be compared with similar F50 values obtained for other explosives or standard primary explosives to obtain an idea of the sensitivity of the explosive under consideration. A higher F50 value indicates lower sensitivity. Variation of this test can be made to obtain the effect of grit on the explosive by mixing the explosive with measured quantities of standardized grit (fine, dry river sand) and subjecting the mix to the test.

Here again most commercial explosives, even cap-sensitive ones, do not show any reaction even when mixed with grit, thus indicating a much lower impact and friction sensitivity as compared to NG-based explosives.

3.4.3
Accelerated Hot Storage (ageing Test)

The commercial AN explosives are not subjected to heat tests for stability as is the case of primary/secondary explosives like NG or PETN.

In case of AN/FO, slurries, and emulsions, AN is by far the largest ingredient in the composition, and the presence of water in the composition brings in different yardsticks of stability.

Stability for these AN explosives in hot conditions of storage means physical stability of either gel/emulsion structure, and for AN/FO that of AN itself. This is done by subjecting the explosive to constant heat at 60 °C in an explosion-proof oven and the physical condition examined at suitable intervals. The explosive can also be tested by firing and VOD measured after it attains room temperature. This test is more of an aging test and is used to predict the long-term stability of the explosive. A correlation between the accelerated storage and normal storage is not exact, but by experience one can come to a reasonable conclusion. For example, 2 weeks in hot storage for water gels/emulsions is approximately equivalent to 3 months normal storage [16]. In normal storage, one introduces variability as conditions of temperature change from season to season and country to country.

The accelerated storage test is used mainly to compare the gel and emulsion stability of these types of explosives especially when new formulations using new thickeners, cross-linking agents, emulsifiers are used. A hot storage temperature of 60 °C (140 °F) is selected as it is known that in many parts of the world magazines in summer especially when they are closed reach this temperature daily for 2–3 months in a year.

3.4.4
Cold Temperature Storage Test

Just as in the hot test explosive is subjected to constant heat, in this test explosive is subjected to storage at constant cold temperature of 0 °C and condition and performance seen at intervals. The explosives are fired only after bringing its temperature up to standard say 21 °C (27 °C for Tropics). This test is not a test for sensitivity to initiation but for examining the physical condition after prolonged cold storage. There are other tests for specifically determining the sensitivity at cold temperature, after cycling between hot and cold temperature, after freezing and thawing cycles, and after cycling at 32 °C. These tests are discussed in detail later.

3.4.5
Thermal Stability Tests Using DTA and TGA Procedures

This gives an idea of the stability of the explosive in an enclosed condition when there is continuous supply of thermal energy (heat). In these tests, the explosive and also AN alone are subjected to differential thermal analysis /thermo gravimetric analysis procedures which are well known and standardized. The exothermicity of the explosive is evaluated from the temperature of initial decomposition and total heat liberated. The effect of impurities, catalysts, new ingredients, metallic powders of different particle size, so on, on the decomposition temperature and rate of decomposition of the explosive can be determined. The possibilities are vast. The data need to be analyzed carefully for its impact on manufacture and usage of the explosive to prevent unsafe situations developing unexpectedly.

3.5
Summary

Testing of explosives both in the initial stages of evolving and later while manufacturing will require different criteria. The test in the initial stages of research and product development needs to give in-depth knowledge of the explosive in all its characteristic. For quality checks in routine production plant and storage, tests need to give only information on the status of the explosive being manufactured and to inform whether it is deviating from the norms set after elaborate study. Above all, the tests need to be simple, cost-effective, and fast so that the plant can correct any defective material being produced immediately before too large a quantity has been produced.

The tests described in this chapter are in a brief manner giving only the basic principle behind the test and the information which can come out of performing such a test and how useful it is to the current products that are subjects of this book, namely AN-based civil commercial explosives. For greater details and a more elaborate list of tests, the author recommends strongly the contents of the book authored by Suc'eska. [5].

In the author's opinion, there is still scope to develop some more tests which are able to come closer in predicting field performance from the test experiment results. The same opinion was expressed as one of the reasons for undertaking a deeper and detailed study [6].

References

1. Cook, M.A. (1974) *The Science of Industrial Explosives*, Ireco Chemicals, Salt Lake City, UT.
2. Mahadevan, E.G. (1976) Recent Developments in Technology of Slurry Explosives, Jahrestag, ICT, Germany.
3. Fordham, S. (1966) *High Explosives, Propellants*, Pergamon Press, Oxford.
4. Mason, C.M. and Aiken, E.G. (1972) *Methods of Evaluating Explosives and Hazardous Materials*, USBM, Bruceton, PA, USA IC8541.
5. Suc'eska, M. (1995) *Test Methods for Explosives*, Springer Publishing, New York.
6. Sarma, K.S. (1994) Models for assessing the Blasting Performance of explosives PhD Thesis. University of Queensland, Australia.
7. Srinivasan, R. and Mahadevan, E.G. (1973) Comparison of strength values of slurry explosives obtained by lead block expansion and balistic mortar methods. *Explosivestoffe*, **5**, 182–185.
8. Dixon, J.W. and Mood, A.M. (1948) A method for obtaining and analyzing sensitivity data. *J. Am. Stat. Assoc.*, **43**, 101–126.
9. Goldstein, S. and Johnson. J.N. (1981) Aquarium tests on aluminized AN/FO. Proceedings of the 7th International Symposium on Detonation, pp. 1016–1023.
10. Lowndes, C.M. and Du Pleiss, M.P. (1984) The double pipe test for commercial explosives. *Propell. Explos. Pyrot.*, **10**, 5–9.
11. Cole, R.H. (1948) *Underwater Explosions*, Princeton University Press, Princeton, NJ.
12. Bjarnholt, G. (1975) Strength Testing of Explosives by Underwater Detonation. Report DS 1977:8, Swedish Detonics Research Foundation, Stockholm.
13. Satyavratan, P.V. and Vedam, R. (1980) Some aspects of underwater testing methods. *Propell. Explos.*, **5**, 62–66.

14. Cameron, A.R. and Torrance, A.C. (1990) The underwater evaluation of the performance of bulk commercial explosives. Proceedings of the 6th Annual Symposium on Explosives and Blasting Research, Orlando, FL.

15. Slykhouse, T.E. (1965) Empirical methods of correlating explosives catering results. Proceedings of the 7th Symposium on Rock Mechanics, USA.

16. Mahadevan, E.G. (1972) Influence of Tropical Storage Conditions on Stability and Performance of High Explosives Especially Slurries, Jahrestag, ICT, Germany.

4
Ammonium Nitrate and AN/FO

4.1
Introduction and History

Any treatise on ammonium nitrate (AN)-based explosives cannot be complete without going into the depth of AN in all the aspects that affect its function as an explosive. AN and AN fuel oil (FO) are inseparable. In order to manufacture acceptable quality ANFO, one needs to have the requisite AN. Understanding how this can be provided is the content of the next few pages. The plethora of information on AN in all its facets is selectively presented keeping in mind the final objective, namely to produce a commercially acceptable ANFO explosive.

AN is one of the most common and commercially important compounds since its discovery in the seventeenth century (1659). The high volumes of AN being produced and used is mainly due to it being the major component in fertilizers and industrial explosives. Serious attempts are being made to introduce AN as a component in solid propellants. If successful, it would further increase demands on AN globally. While the role of AN as a fertilizer is to provide N_2 both in the form of NH_3 and nitrate ion, its role in the explosive is to function as an oxidizer to provide a cheap source of oxygen in the explosive to combine with the fuel and release energy. That AN itself possesses explosive character became apparent in a tragic way through disasters involving AN alone in explosions of severe magnitude. The most famous and well-known ones are the accidents in Upper Silesia (1921), Oppau, Germany (1921), Brest, France (1947), Texas City, USA (1947), Toulose, France (2001). A thorough investigation by scientists of its decomposition chemistry, thermochemistry, crystalline behavior, physical properties, heat of formation, and oxygen content produced enormous and satisfying data to enable solutions to be found for problems cropping up in its industrial applications such as physical strength, porosity, caking, release of energy, and sensitivity to initiation. An important conclusion reached in the early 1950s was that while AN by itself possessed explosive character and under certain conditions can function as an explosive albeit inefficiently, it could be made much more effective as an explosive by combining it with a small amount of an efficient fuel. Thus ammonium nitrate fuel oil explosive was born.

Ammonium Nitrate Explosives for Civil Applications: Slurries, Emulsions and Ammonium Nitrate Fuel Oils,
First Edition. E.G. Mahadevan.
© 2013 Wiley-VCH Verlag GmbH & Co. KGaA. Published 2013 by Wiley-VCH Verlag GmbH & Co. KGaA.

4.2
Physical and Chemical Properties of Ammonium Nitrate

4.2.1
Basic Data

Literature gives enormous data on the above, but one of the most comprehensive single source is the paper authored by Oommen and Jain [1].

AN in its usual form, unless subjected to special processing, exists as a crystalline hygroscopic solid, melting point (MP) around 169.6 °C. Given below are some of the values for important properties (Table 4.1).

4.2.2
Decomposition Chemistry of AN

While pure AN is stable at room temperature, on heating in the open melts at 169 °C, boils at 210 °C, decomposes around 230 °C, and deflagrates at above 325 °C. When confined and heated, AN can explode beyond 260 °C depending on the rate of heating.

Decomposition chemistry of any substance is the key to its explosive behavior and accordingly that of AN has been extensively studied and published. When analyzing the experimental data, one needs to pay attention to underlying factors like rate of heating, pressure, and purity as these considerably influence the results. Even though several reaction pathways are suggested, no single mechanism is existing that can explain all the aspects of AN decomposition. Still there is a fair degree of acceptance of a specific reaction mechanism at a specific temperature [1, 2].

For the purpose of understanding the usefulness and the hazards associated with use of AN in AN explosives, especially ANFO, it will be sufficient to know that rate of heating and confinement are two very important parameters influencing the decomposition route and end-product generation during AN decomposition. It is suggested that while high rate of heating encourages surface decomposition which is endothermic and produces NH_3 and HNO_3, lower rate encourages bulk decomposition which is exothermic and produces N_2O and H_2O. The presence of impurities, which may act as catalysts are many [2]. Other additions may include metallic fuels which enhance the heat energy liberated during decomposition. It is this phenomenon that is used in formulating ANFO explosive which gives better blast performance at higher bulk densities. Modes of thermal decomposition suggested for pure AN are presented in Ref. [3].

At 180 °C	$NH_4NO_3 = NH_3 + HNO_3$	-10 kcal/mol
At 250 °C	$NH_4NO_3 = N_2O + 2 H_2O$	$+10$ kcal/mol
At > 300 °C	$2 NH_4NO_3 = 2N_2 + 4H_2O + O_2$	$+28$ kcal/mol

Under confinement detonation occurs as

$$4NH_4NO_3 = 2NO_2 + 8H_2O + 3N_2 \quad + 27 \, kcal/mol$$

With just enough fuel added to consume all the available oxygen from AN, the reaction is

$$3NH_4NO_3 + CH_2 = 3N_2 + 7H_2O + CO_2 \quad + 82 \, kcal/mol$$

There is almost threefold increase in the heat energy released as compared to that where AN rapidly decomposes without the presence of fuel.

4.2.3
Phase Transition in AN and its Importance in Explosives

AN crystals exist in different forms at different temperatures. The change in the crystal structure with temperature is known as *phase transitions* and such changes are also accompanied by change in the physical properties such as volume, size of the crystal, and mechanical strength.

Phase transition of AN is one of the most studied subject of inorganic chemistry and extensive information is available in literature [1, 2]. The main reason for such a lot of study is the fact that the phenomenon accompanying phase transitions adversely affects the desirable properties of AN used in the explosive and fertilizer industry. Caking and loss of flowability have been found due to phase changes as a result of temperature cycling in storage. There is also loss of sensitivity to initiation in explosives. At the same time, caked AN has detonated resulting in huge loss of life when attempts were made to break the caked mass.

Phase changes detrimental to the commercial usage of AN are seen not only in crystalline AN but also in spherical AN (prills), which is now the most commonly used physical form in industries. A complete description of crystal structure and crystallographic data and literature survey on this subject can be found in [1]. The following are the temperatures of the phase transitions:

Table 4.1 Important properties of AN.

Molecular weight	80
Heat of formation	1098 cal/g
Heat of explosion	346 cal/g
Density	1.725 g/cc
Solubility in water at $20\,^\circ C$	66/100 g
Available oxygen	20%
Estimated flame temperature	$1500\,^\circ C$

- V at $-18\,^\circ$C to IV
- IV at $32\,^\circ$C to III
- III at $84\,^\circ$C to II
- II at $125\,^\circ$C to I.

The transition of greatest importance for practical purposes is the one occurring at $32\,^\circ$C where the tetragonal structure passes into rhombic and vice versa. In most manufacturing and storage conditions, the AN undergoes cycling around $32\,^\circ$C, sometimes almost daily for several weeks in a year. In very cold countries in winter, the V to IV transition can also be a factor. The volume change is 3.7% for III–IV and 2.8% for V to IV. In literature, many ranges have been reported for the transition temperatures due to influence of moisture, impurities, method of crystallization, prilling, and thermal history of the sample. A wide range of experimental techniques such as X-ray diffraction, Differential Thermal Analysis (DTA), differential scanning calorimetry (DSC), neutron diffraction, and dielectric measurements have been used, but still there is some inconsistency as to the results, especially if very precise and clear-cut temperatures of transition are being sought (for most application $32\,^\circ$C is taken as the benchmark for IV to III transition). The variations and inconsistencies observed in the transition temperatures are further compounded by the complex phenomenon of hysteresis observed during cycling (heating and cooling) of AN. This behavior is also dependent on moisture content, rate of heating and cooling, and the period at which the AN remained above or below the transition temperature. Crystallographic studies through X-ray diffraction have given insight into the crystal structure of different phases and pointed out the route by which phase transition temperature can be possibly manipulated. Based on the crystal structure of the phases IV and III and their differences, it was thought that altering the crystal structure in favor of one or the other could be achieved by altering the phase transition temperature. As IV to III change at $32\,^\circ$C was of greatest practical importance, much work was done on finding additives to shift and eliminate the transition temperature to avoid the ill effects of cycling through $32\,^\circ$C on daily basis. Although much information is available in books, scientific papers, patents, and commercial literature, the exact percentage and nature of additives used by the AN manufacturers may not be apparent. For end users of AN, the behavior of the AN in storage is more critical. Of course claims of the manufacturer regarding phase stabilization or shift and elimination of a specific transition temperature could always be checked by running DTA/DSC scan and comparing it with the thermogram of pure AN. The following additives are quoted as being beneficial to the explosives in terms of storage properties and stability. Most prominent are $-KNO_3$, alkali metal nitrates, KF, KCl, K_2SO_4, $K_2Cr_2O_7$, $Cu(NO_3)_2$, $Mg(NO_3)_2$, $Zn(NO_3)_2$, $(NH_4)_2SO_4$, $Na_2B_4O_7$, metal dinitramides, and metal oxides. The criteria for choosing the right additive are the ease of incorporation in the process used for AN manufacture, environmental considerations, energy contribution, cost, and above all effectiveness under most storage conditions that are usually found in industrial operations.

4.3
Manufacture of Ammonium Nitrate

Chemically AN can be made starting from the basic elements N_2 and H_2:

$$N_2 + 3H_2 = 2NH_3$$
$$NH_3 + 2O_2 = HNO_3 + H_2O$$
$$NH_3 + HNO_3 = NH_4NO_3$$

The above equations are simplified representation of the process. In practice, oxidation of ammonia can also produce oxides of nitrogen which have to be disposed off in a nonpolluting manner. The final product (AN solution) will have water which will need to be removed.

There are many entry points in the production cycle depending on the need to integrate backward, availability of capital, and final economics. Most AN production starts from NH_3 and produces HNO_3, which then absorbs NH_3 to produce AN. One could also buy both NH_3 and HNO_3 from outside and implement only the final step. But for ease of handling and energy balance, producing HNO_3 on-site is preferred. Once the absorption of NH_3 in HNO_3 is complete, the solution of NH_4NO_3 containing varying amounts of H_2O depending upon the strength of HNO_3 used is subjected to different types of unit operations of evaporation drying (removal of H_2O), crystallization, and prilling to obtain AN in different physical forms and moisture content. In some instances, the AN liquor with water content of 20% or 7% is removed for supply to manufactures of slurry and emulsion explosives as AN melt which they can directly use to make the oxidizer blend.

Historically AN was produced in large commercial quantities in crystalline form till 1955 when prilling process was established. The crystalline AN was mainly used as fertilizers and also in nitroglycerine (NG)-based dynamites as oxidizer and diluents. Use of crystalline AN in NG-based explosives gave rise to a series of explosives called *gelatins*, which were very popular civil explosives as they were both cap-sensitive, had adequate power, and were reasonably safe for manufacture and handling. The addition of AN brought down the NG content from around 60% to 32% initially in gelatins and later down to 15–17% in semigelatins.

The increased use of crystalline AN brought its own problems as the AN used was prone to caking due to temperature cycling around $32\,°C$, which we have discussed earlier. Volume change and fineness brought about by cycling coupled with the moisture inherent in the crystalline AN contributed to caking. The fine AN formed rigid bridges on pressure from contact with other particles. Also the saturated droplets of AN on loss of water formed rigid structures with the result that the AN, even though well mixed with NG, formed hardened mass and got deformed into shapes other than cylindrical. These deformed cartridges presented loading difficulties in boreholes and some accidents have been reported when efforts were made to bring them back to original cartridge shape by applying physical force.

In the early 1950s, crystalline AN was also used in making powder explosives mixed with Di Nitro toluene (DNT)/Mono Nitro toluene (MNT) and FO. These explosives were non-cap-sensitive and were used only in large boreholes. The caking phenomenon was prevalent even here and its effect on the performance of the explosive was much worse than with NG-based gelatins and frequent blast failures were reported. These explosives had the cost advantage over the gelatins and when freshly made and used soon thereafter could give adequate performance in very large diameters. The crystalline AN used as fertilizer while effective as N_2-fixer caused great difficulty for transporting and spreading uniformly in the field due to caking and lumping. Physically breaking down such lumps was time-consuming and dangerous as evidenced by the serious explosions which took place in Texas and other locations mentioned in literature as examples.

The characteristics of one of well-used AN crystallized products manufactured in Sweden are given below.

NH_4O_3	99.5%	Moisture	0.2%	Coating	0.1%
Bulk density	1.1 kg/l	Crystal size	0.25–1.0 mm	—	92%

The phenomenon of caking in crystalline AN was of sufficiently serious magnitude to come to the attention of scientists and engineers and motivate them to find ways and means to overcome the problem. Physically coating the individual crystals to keep them separate was tried out with limited success. Some of these coatings were inert and took away the energy of the explosive. Some were organic adding to the sensitivity and introducing an enhanced hazard situation. The most optimum results obtained involved use of stearic acid and its salts as also long chain fatty acid amines in very small quantities like 0.5–1.0%. Use of azo dyes (water soluble) such as acid magenta [3] during crystallization was also not found very successful, and serious thoughts were then given to modify the 32 °C transition temperature to reduce phase IV to III transition during normal storage as well as decreasing surface contact between AN particles by introducing spherical structure. These spherical AN came to be known as *prilled AN* and the process of producing such AN as prilling.

4.3.1
Prilled Ammonium Nitrate

Prilled AN was developed during the late 1940s after the World War II as demand for fertilizers increased enormously, and countries like the USA where fertilizer program was based on use of AN was the drive engine. The criteria for a suitable AN fertilizer were cost, ease of production, transportation, and distribution in the farmland. Granulation was attempted by varying the crystallization process conditions being used then, but soon prilling was the preferred process as the product obtained (prills) overcame many of the problems encountered earlier with AN in crystalline form.

Table 4.2 Properties of Prilled AN (Explosives Grade).

Properties	Type 1	Type 2
Bulk density	0.79–0.83 kg/l	0.85
Moisture	0.1%	0.15%
Oil absorption	8.0%	<9.0% >7.5%
Inert coating	0.0%	—
Organic coating	0.1%	0.1%
Size 1–2 mm	80%	—
Size >2 mm	17%	—
Size 1.6–2.4 mm	—	98%

Initially the prills produced were the so-called high-density (HD) prills. These were having inert coating to prevent strong agglomeration, hard surface for transporting without damage, but had poor explosive functionality. By and large, these HD prills were used as fertilizers. In the meanwhile to satisfy the requirements of the explosive industry, which had discovered AN/FO as a substitute for NG explosives in large-diameter blasting, process parameters were altered to obtain lower density (LD) porous prills. The process parameters involved evaporating the moisture or water content of the AN before prilling so as to develop capillaries on the surface of the prill, regulating the rate of cooling and droplet size of AN solution as it is being sprayed in the prilling tower, prilling tower height, addition of specialty chemicals for suppressing or modifying 32 °C phase transition, and preventing volume changes during temperature cycling (Figure 4.7). Inert coatings effective at low percentages were added after prilling. A large amount of applied R&D succeeded in developing processes which could yield the desired quality AN prills for the explosives industry. Patent literature in plenty claimed many innovations to achieve the desired results such as low height prilling tower, several stages of cooling for hardening and for developing capillaries on the surface of the prill, fixed/rotary dispersion of AN liquor at the top of the prilling tower, and so on. The current status is that a process with an ability to create a narrow prill size distribution and uniform oil absorbency has become available from most process developers. Table 4.2 shows the specification of two standard AN prills.

Notwithstanding what has been said above, these are still many forms of AN prills produced being not up to required minimum standards and correspondingly the AN/FO made using such substandard prills give poor blasting efficiencies.

The basic prilling process consists of forming an AN liquor of desired concentration by neutralizing fuming HNO_3 with NH_3 and evaporating most of the excess water formed. The AN liquor is then taken up to a prilling tower (60 m height and 5–8 m diameter) and allowed to drop down the tower by gravity. The droplets

Table 4.3 Comparison of HD and LDAN prill properties.

Properties	High density	Low density
Physical form	Hard spheres	Porous spheres
Inert coating (%)	3–5	0.5–1.0
Bulk density (kg/l)	0.90–0.98	0.70–0.82
Free moisture (%)	0.15–0.6	0.1
Oil absorbency (%)	3–4	>7.0
Porosity (%)	<5.0	15–22
Crushing strength (kg)	2.5–4.0	1.4–1.8

Table 4.4 Properties of extra porous AN prills.

NH_4NO_3 purity	98.6%
Coating	<0.8%
Oil absorption	>11%
Moisture	<0.2%
Bulk density	0.67
Porosity	29%

get cooled and the remaining water removed by a countercurrent of cold, dry air blown upward from the bottom of the prilling tower. Spherical particles with a hard outside surface, which are also porous, are formed. The prills are classified into fractions by size, cooled, coated, and bagged. Additives for phase stabilization are added before prilling. Engineers skilled in the art of prilling are able to tailor-make the prills as per requirement of oil absorbency, crushing strength, sphericity, uniformity in particle size, and so on, by varying the parameters of the prilling process. Table 4.3 shows comparison in the properties of HD and LD prills.

Properties of the extra porous AN prills are given in Table 4.4.

The AN/FO explosives made with the extra porous prills are claimed to show 10–15% increase in velocity of detonation (VOD) and other explosive properties compared to AN/FO made from standard porous prills. There is another commercially available unique product claimed by the manufacturer of imparting enhanced explosive performance when used in AN/FO in spite of the lower bulk density. These special type prills have hollow polymeric microspheres incorporated during the prilling process and distributed uniformly throughout the mass of AN prills. These entrained microspheres bring in additional hotspots in the body of the AN, and consequently when such AN is used for making AN/FO the explosive properties and field performance exhibited by such an AN/FO are claimed to be significantly superior to that shown by AN/FO where standard porous prills have been used.

4.4
Ammonium Nitrate Fuel Oil Explosives

4.4.1
Background

At this juncture, it may be a good idea to go into the world of AN/FO. It is the most basic of the AN civil explosives. As the name implies, it is a mixture of AN and FO. AN is the oxidizer and FO forms the fuel component. The oxygen balance is zero at 5.7% of FO to 94.3% AN. Output of energy from AN/FO is maximum at this composition. We shall discuss later the effect of lower or higher fuel content and the characteristic of AN itself on the explosive properties.

As mentioned earlier, initially the HD fertilizer-grade AN prills were used in AN/FO due to nonavailability of explosive-grade prills and poor blasting results were often obtained. In fact below 5-in.-diameter boreholes, there were many failures. The situation changed drastically with the availability of porous low-density prills. The latter have now become standard major ingredient of AN/FO. The improved performance when using LD porous prills is attributed to the increased number of hotspots available in the Low Density Porous Prilled Ammonium Nitrate (LDPPAN) due to its microporous structure of air voids enclosed within the body of the prills.

Thus the performance of AN/FO is only as good as the AN prills used and the efficiency of mixing with the FO. Influence and effect of both these parameters on the performance of AN/FO explosive are discussed in detail later.

4.4.2
AN/FO Manufacture

The quantities being manufactured are huge and run annually into couple of million tonnes. Individual plants are set up very near the mine sites to enable less storage and speedy delivery to the end user.

4.4.2.1 **Mixing Process and Equipment**
The operation is a solid/liquid mixing with other solids' addition subsequently. The latter additions are of small percentages. The initial mixing is invariably between AN and FO. Mixing can be either in a batch process or continuous. The latter is preferred for ease of handling.

Horizontal Batch Mixing   In the batch process, the AN prills are taken into a standard, polished, stainless steel, or aluminum double helix mixer, equipped with forward and reverse possibilities. The clearance is at least $^3/_4$ in. between the blade and the walls of the mixer. The shaft is suspended from outside through a gland packing. Vents are provided from the sides of the mixer to spray in the FO. The mixture after process is complete can be pushed out from a centrally located exit operated from outside. Batch size is usually 500–600 kg.

Mixing time is around 12–15 min. FO is sprayed in with sufficient speed that the entire quantity is over in 5–8 min. Sometimes the FO contains an oil-soluble dye of orange or red color to give a visual indication of the uniformity of mixing. After the mixing process with FO is complete, other additions are made such as Al, Fe powder, and water-proofing agents. The final mix is un-loaded into a hopper from where the material is moved by screw feeder into packages, which are either polythene cartridges clipped at one end which will assume on filling the required diameter/weight or bulk paper bags laminated with polythene film. Quantities of individual packages could range from 2.5 to 25 kg.

Double-cone blenders are also used as they are efficient in mixing with no moving parts such as a mixing blade inside making it safer if any foreign bodies inadvertently get into the mixer. Generally the prills are screened before being added to the blender. Arrangements are made for spraying in the oil as the mixer is being rotated. Discharge from a double-cone blender is easier than horizontal blender. Other steps in the process leading to finished packages are the same as mentioned earlier.

Nauta mixers are also used which are very efficient. My experience is that there are greater chances of fines being produced by this method.

4.4.2.2 Continuous Process

Currently the most preferred method of making AN/FO is by a continuous process using screw conveyors. Right from unloaded AN prills to storage or bagging the material is continuously flowing inside screw conveyors. At the start itself, FO is sprayed onto the AN and gets mixed as the AN moves along and turns over. The rate of FO addition is fixed in relation to the speed at which AN is being conveyed so that the final mix contains the desired ratio of AN/FO (94/6). Colored dye is used in the FO for visual check regarding dispersion uniformity. The length of the conveyor is adjusted to ensure adequate mixing but not so much as to physically damage the AN prill structure and produce fines as all screw conveyors do exert a certain amount of mechanical pressure on the material being conveyed.

Rates of 500 kg/min are quite common. The advantages of these continuous plants are that they can be started and stopped easily, need very little man power to operate, and the inventory of finished explosive in storage can be kept to a minimum especially if there is continuous loading on to bulk delivery vehicles. The material (explosive) is generally quite fresh after mixing and should deliver optimum performance. Safety features can be easily built in for monitoring hazards.

4.4.2.3 Bulk Delivery Systems

The stationary continuous process of mixing AN prills with FO described above is adopted for bulk delivery also. A mini mixing plant is mounted on the chassis of a truck. Power take off from the truck's engine delivers the necessary power for mixing AN/FO stored separately on the truck. After mixing on the truck, the AN/FO is straightaway augered down the hole. Once the borehole is filled, the

Figure 4.1 AN/FO Bulk Delivery Vehicle.

vehicle moves on to the next and repeats the process. Delivery is from the side in most Bulk Density (BD) vehicles (Figure 4.1) so that a watch can be kept easily by supervisory personnel and sample of the mix going down can also be collected easily. Up to 500 kg/min delivery rates for large-diameter boreholes (7–12 in.) are attainable as long as there is free flow of the prills. Some BD vehicles are nothing but pure delivery vehicles and there is no mixing of AN with FO on the truck. This is the simplest system in that the already mixed AN/FO in a stationery plant is loaded on to the truck which goes on to the mine face and augurs the ready mix down the hole. In this instance, the delivery vehicle is treated as an explosive-carrying van and is subject to such rules that govern this category of vehicle transport. This kind of truck is also the simplest and cheapest but the efficiency of the delivery would depend on the free flow of the mix, which can cause problems of bridging and clogging if the mix has been stored under pressure for a period of time before being loaded onto the bulk delivery vehicle.

4.4.3
Properties of AN/FO

4.4.3.1 **Physical**
AN/FO made out of AN prills is a free-flowing loose solid capable of functioning as an explosive under certain conditions of boostering, confinement, and diameter of charge. AN/FO made out of HD and LD prills do show differences in their VOD and initiation sensitivity (Table 4.5).

It can be seen from this data that use of LDPP is the only way to obtain acceptable performance in AN/FO.

4.4.3.2 **Oil Absorbency and Porosity/Bulk Density/Crushing Strength**
These are important properties directly affecting the explosive performance in AN/FO.

Table 4.5 Detonation properties of AN/FO made with high- and low-density AN prills.

	AN/FO made with high-density prills	AN/FO made with low-density porous prills
Unconfined minimum diameter for detonation (in.)	9″	2 1/2″
Confined minimum diameter for detonation (in.)	4″	1 3/4″
VOD in 4 in. (m/s) open	1500–1800	4000
Inert material (%)	3–5	0.1–0.5
Fume characteristics	Very bad	Acceptable
Oxygen balance	Cannot be balanced by FO alone	Can be balanced by FO
Oil absorption of AN prill used	3%	8% and above
Air voids (hot spots)	Negligible	Considerable

The surface of AN prills made for the use of explosive industry is full of capillaries leading into the body of the prills. There are also voids entrapped inside the prill. Both these are achieved during removal of water in the stages of prilling and drying. Obviously the more the number of voids/capillaries per surface area of the prill, the greater is its porosity and its capacity to absorb and retain FO inside its body.

A simple and fairly accurate method of finding oil absorbency of prilled AN is to take a known weight of the prills, say 100 g, and immerse it fully under FO No. 2 (standard) for 1 h after which the excess oil is poured out. The wet AN prills are then taken out, spread on a blotting paper sheet, and using absorbent paper the excess oil on the surface of the prill is removed. Care is taken to see that the oil is not wicked out from inside of the prill. This will happen if the absorbent paper is fresh and held pressed for too long on the prill. Once the surface excess oil is removed, the AN prills are weighed and the gain in weight is due to the oil absorbed and held within the prills. It is expressed as percentage or grams per cubic centimeter.

Another method involves dissolving a known weight of AN after mixing with FO in water and measuring the volume of oil liberated. Knowing the density of the oil used, one can arrive at the weight of oil coming out of a known weight of AN/dissolved and hence percentage of oil absorbency can be calculated.

Porosity of a prill is the free space in percentage within itself. A good porous prill has around 21% porosity.

Bulk density is inversely proportional to porosity. At constant porosity, bulk density depends on size and sphericity of the prills. Bulk density is important in blasting as it is a measure of the energy available for a fixed volume of loading. Higher bulk density is preferred as it concentrates energy in a smaller volume and is useful to remove the bottom toe in blasting. However increased density *per se* reduces the voids in the prills and affects the sensitivity of the prill to initiating shock wave due to the lesser number of hotspots available.

A rough method of checking the uniformity of oil absorption is a visual check of the coloration on the prills when mixed with FO containing a dye (orange or red).

A very high porosity and oil absorbency >12% sometimes lead to uneven absorption as the rate of absorption is very high and if the oil is not sprayed in a fine mist, pockets of high and low oil content result in the AN/FO. This is not desirable from blasting performance as explained later.

Crushing strength is another important property of the prill giving an indication of its fragility to mechanical impact and pressure usually experienced while transported in bulk. For end user, it is important to get the AN prills in its original physical form to ensure uniformity in the loading density and reduce wastage in the form of fines. While a certain amount of fines are acceptable in that they increase the loading density by occupying space between prills, there is a limit since increase in overall density beyond 1.0 g/cc will bring in insensitivity to detonation which is to be avoided at all cost in a loaded borehole (Figure 4.2).

Crushing strength is measured by applying a fixed load on the prills contained in a cylinder for a determined length of time after which by sieve analysis amount of fines formed is measured, which quantifies the crushing that has taken place. Generally the method is used for comparison. Fertilizer-grade HD prills can be the standard.

A crushing strength number between 0.14 and 0.35 kg/cm^2 ensures adequate mechanical strength and porosity. Crushing strength and porosity are properties inversely related to each other. A prill with low porosity has high crushing strength and vice versa. This relationship is not hard to understand when one looks at the structure of low- and high-porosity prills. Thus it is a balance against each others' beneficial effect. The only exception is when the product is machine loaded into

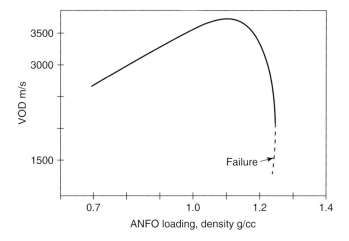

Figure 4.2 Effect of density on detonation of AN/FO.

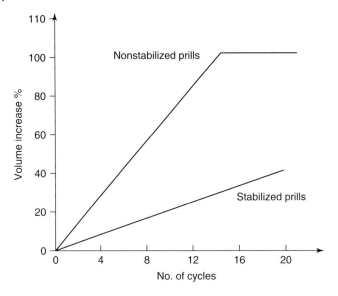

Figure 4.3 Effect of cycling of AN.

small boreholes where the prills should be fragile enough to break down in the borehole due to air pressure and porous enough to be sensitive to initiation even at higher than original prill density since a certain amount of compaction would have taken place in the boreholes. For such application, specially made extra LDPP are available.

4.4.3.3 Resistance to Effect of Temperature Cycling

As mentioned earlier in detail, cycling around 32 °C is the one which affects properties of AN prills in storage.

The AN prills on being subjected to phase transformation from IV to III undergo volume change (Figure 4.3) leading to fineness due to reduction in crushing strength. Further if moisture is also present, the AN prills will also tend to cake into lumps due to bridging of the fines on absorption of water from the atmosphere.

Method of temperature cycling, which claims to stimulate actual storage conditions in tropical climate, is described below.

A known volume of AN prills after removing fines is kept in a graduated measuring cylinder and the level noted. The cylinder is kept in an oven and subjected to hot period (day time) of 8 h at 50 °C and a cold period (night time) of 16 h at 20 °C (cooling facility necessary if ambient temperature does not go below 25 °C). One cycle has 8 h of hot period and 16 h of cold period. Five such cycles are performed continuously. At the end of five cycles, the cylinder with AN is taken out gently and the volume occupied is measured. Next the fines quantity is determined by screening. Fines are those less than 1 mm in size.

4.4.4
Characteristics of ANFO

The important properties of AN/FO and their role in affecting blasting performance is of great interest to both the manufacturer and the end user as well.

4.4.4.1 **Density/Strength**

Density by definition in its classical sense is mass (weight)/volume. In case of an explosive, density will determine the volume (height) occupied in a borehole by a given weight of the explosive if the diameter of the borehole is known. Two explosives of same density and weight will occupy the same volume in a borehole. The blast performance will then depend on the composition of the explosive and its VOD. On the other hand for an explosive of higher density to occupy the same height in a borehole, more of the explosive has to be loaded and this means more energy can be placed per unit volume. The greater this is, the more advantageous it is for blasting.

In case of AN/FO, depending on the prill type the bulk density ranges from 0.8 to 1.1 g/cc. The lower range is for poured or bulk loaded and the higher range for packaged product. Metallic additives increase the density of pure AN/FO by 15–20%. The density of AN/FO cannot be increased beyond a certain value as the explosive will tend not to initiate because of lower response to boostering (detonation wave) at the higher density. An optimum has to be struck between density and sensitivity. Density affects critical diameter values also. At constant density, the VOD increases initially with diameter but tapers off to a constant value (Figure 4.4).

Cup density measurements are standard methods of measuring the density. In a metallic cylinder of known volume, AN/FO is poured in till it is fully filled. The weight of cylinder before and after filling is used to calculate cup density (poured). A variation of this is to tap the cylinder while filling so that no voids due to bridging influence the result. By this means more of the AN/FO goes into the same volume. The density calculated using these values are known as *cup density* (tapped).

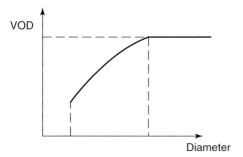

Figure 4.4 Diameter versus VOD in AN/FO.

4.4.4.2 Strength of the AN/FO Explosive

It is best to use underwater energy measurements to calculate the strength of AN/FO. Detonation pressure can be calculated using measured VOD values. AN/FO is one of those explosives where despite having low-detonation pressure, the performance in the field in certain conditions is very good mainly due to the high gas volume it generates. This is due to the very high content of AN (>90%) in its composition.

Total gas generated by AN/FO at 25 °C is 1050–1100 l/kg. Blasting gelatin generates around 800 l/kg.

In many instances, historically AN/FO is taken as the standard and other explosives compared with it. Shock and bubble energy values of AN/FO obtained from underwater measurements have been used for obtaining optimum blast designs useful for actual field applications.

A summary of important properties to be kept in mind while dealing with AN/FO are density, primer size and weight, unconfined critical diameter, VOD at different diameters, weight and bulk strength, ideal detonation velocity, fume character, water proofness, heat of explosion, and explosion temperature.

4.4.4.3 Energy Content of AN/FO

AN, which is the major component in AN/FO, undergoes exothermic decomposition in many ways as mentioned in the earlier chapter. During detonation, the reaction to be considered are

$$NH_4NO_3 = N_2 + 2H_2O + \tfrac{1}{2}\ O_2 + 1440 \text{ kJ/kg}$$

When fuel of the right quantity is present to consume all the oxygen liberated (oxygen balanced composition of 94.3% AN + 5.7% FO), the reaction can be represented as

$$3NH_4NO_3 + CH_2 = N_2 + CO_2 + 7H_2O \quad + 3900 \text{ kJ/kg}$$

There is almost a threefold increase in energy released when AN decomposes in the presence of the right amount of FO.

The energy released decreases drastically to 2500 kJ/kg if the fuel provided is not sufficient to consume all the oxygen released by the decomposition of AN.

$$5NH_4NO_3 + CH_2 = 4N_2 + CO_2 + 11H_2O + 2NO \quad + 2500 \text{ kJ/kg}$$

96.6 % AN + 3.2 % FO

When overfueled, the reaction path is

$$2NH_4NO_3 + CH_2 = 2N_2 + CO + 5H_2O \quad + 3400 \text{ kJ/kg}$$

92 %AN + 8 %FO

It can be seen that lower than 5.7% oil content brings down the energy released considerably whereas slight excess of FO has a smaller reduction. Thus it is prudent to keep the oil content slightly higher than the ideal, say 6–6.5% (Figures 4.5 and 4.6).

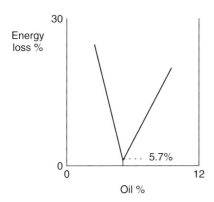

Figure 4.5 Effect of oxygen balance on energy loss in AN/FO.

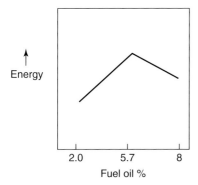

Figure 4.6 Energy output versus fuel content (AN/FO).

4.4.4.4 Velocity of Detonation and Effective Priming

It is seen that bulk density as well as charge diameter have an influence on the VOD values. The VOD is at a maximum when the energy release is completed in the actual detonation zone. If part of the energy release takes place in the reaction front traveling behind the detonation zone, the VOD will be less than ideal. The peak detonation pressure will be lower but the duration is longer. As the diameter of the charge increases for the AN/FO even with differing densities, the measured VOD approaches the ideal, around 4600 m/s (Figure 4.7).

The VOD increases with charge diameter but at levels of a constant value. There is a measurable difference at small diameters but above 4 in. diameter most of the AN/FO, if made properly from standard LDPP, have same VOD confined and unconfined. More porous AN prills do exhibit greater sensitivity to priming and higher VOD even in the 40 m to 60 mm diameter range. VOD also depends on AN prill size. Greater the prill size, lower the VOD.

Priming and Boostering In most blasting operations, AN/FO is used in medium/large diameter boreholes and requires a booster for initiation. Only in

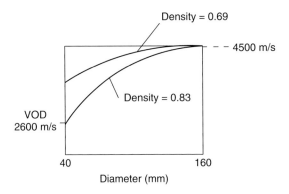

Figure 4.7 Effect of density of AN prills on VOD of AN/FO.

some special applications where small-diameter blasting is resorted to initiation can be by a commercially available detonator. Sometimes a specially made stronger detonator is used to be absolutely sure of initiation.

To obtain the maximum benefit of energy release from AN/FO and to sustain propagation in the column, the AN/FO must be made to attain its steady-state velocity soonest and be made to maintain it all along the column. For this shock wave of sufficient intensity (detonation pressure) and duration applied across the cross-sectional area of the explosive charge needs to be generated in close proximity (contact) with the AN/FO. Preferred choice is pentolite (cast booster explosive of 50% pentaerythritol tetranitrate (PETN) and 50% trinitrotoluene (TNT)) or a High Density (HD), high VOD, cap-sensitive explosive of the same diameter. The detonation pressure generated by the booster should be higher than that of the AN/FO. Priming of AN/FO requires care; otherwise it can lead to disappointing results. The priming requirements are very dependent on the condition of the AN/FO itself as well as external factors such as borehole diameter, borehole condition, degree of confinement, and AN prill characteristics. Booster used for priming AN/FO in small diameters may be unsuitable for priming AN/FO in large diameters. While in theory a single primer should be adequate to initiate steady-state velocity in AN/FO, the explosive by itself is expected to sustain it throughout the column. Borehole VOD measurements at the bottom and top of the column have shown that it is possible if all factors are in favor of preventing losses of energy from the detonation front. But in practice to ensure no failure occurs in propagation, primers are placed in the column also at a distance of 2–3 m depending on the conditions. As a percentage of boostering in small diameters it can be as high as 10%, whereas in large diameters it can be as low as 0.2%. It is accepted that reducing amount and cost of priming and causing substandard blast effect are false economies. On the other hand, overboostering does not make sense either. In explosive chemistry, AN/FO cannot release more than theoretical or ideal amount of energy it contains as per its composition however high the initiating shock wave pressure. In the borehole for a small distance beyond the primer, the VOD can be over the practical VOD of AN/FO at that point but very soon the

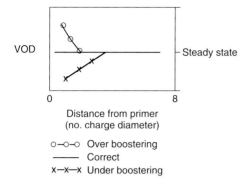

o—o—o Over boostering
——— Correct
x—x—x Under boostering

Figure 4.8 Effect of primer on AN/FO.

VOD falls back to the steady-state VOD of AN/FO which is lower (Figure 4.8). Overboostering will not only add to the cost but may lead to the phenomenon of dead pressing that results in slower energy release by the AN/FO.

4.4.4.5 Mechanism of Detonation Propagation in AN/FO

Measured detonation velocity in most AN/FO is lower than ideal. It is only under confinement and in very large diameters it approaches close to ideal. The mechanism of propagation through hotspots generated by adiabatic compression of air enclosed in the porous structure of the AN prill has found some acceptance as measured values of VOD are higher for AN/FO made with more porous AN. It is suggested that if all the pores are not fully occupied by FO achieved by addition of less quantity of oil than the absorption capability of the prill (6% against say 11%), the voids are acting as source of hotspots and the detonation velocity is enhanced. In one instance, microballoons (enclosed glass bubbles) have been incorporated while prilling AN. In another instance, extra porous AN/prills have been produced by special process conditions and additives.

A comparison of the rate of detonation when using such AN in AN/FO is given in Table 4.6.

Table 4.6 Effect of Porosity of AN on VOD of AN/FO

Diameter (mm)	Standard prills VOD (m/s) in steel pipes	Extra porous prills VOD (m/s) in steel pipes
50	3000	3500
127	3950	4250
250	4300	4600
	VOD (m/s) in the open	VOD (m/s) in the open
88	2600	3050
		3375
Bulk density (g/cc)	0.85	0.70

However due to the diameter effect, all these varieties of AN/FO do not show significant difference in VOD beyond 125 mm and hence the standard AN with higher bulk density is for larger diameters preferred (Figure 4.8).

4.4.4.6 Influence of Fuel

We have already seen that for the best results, 5.7% fuel content is required. Commensurate with safety consideration, FO with flash point above 60 °C is to be used. It was also seen that more viscous oils tended to lower the VOD of AN/FO and mixing is also not perfectly uniform. Other criteria for a good fuel are high thermal energy when oxidized, ability to react fast, easy availability, and cost.

Diesel oil (boiling point 200–250 °C) is now taken as the benchmark fuel for AN/FO. Lowering the flash point increases the VOD of the AN/FO, but correspondingly reduces the safety in handling and increases fire hazard.

4.4.4.7 Effect of Moisture/Wet Boreholes/Water-Resistant AN/FO

The moisture content in AN prills used for making AN/FO has deleterious influence on its long-term thermal stability and caking properties. Manufacturers of Low Density Ammonium Nitrate (LDAN) prills tend to keep the moisture content of the finished product at less than 0.2%. The effect of moisture or water in the borehole has a more immediate dramatic effect on the performance of the AN/FO. The VOD of the explosive drops depending on its absorption of the water in the borehole, which in turn depends on the time it spends in contact with the water and whether the water is static or flowing. The relationship between VOD and moisture of AN/FO experimentally determined in steel tube confinement is shown in Figure 4.9. Behavior in a borehole could be worse as the confinement is not absolute and ideal.

Thus as a prudent precaution, AN/FO is not used as such in watery boreholes and this is one of the most serious drawbacks of this explosive. In many instances, it may still be practical and economical to pump out the water, line the borehole

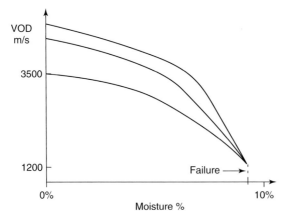

Figure 4.9 Effect of moisture on VOD of AN/FO.

with a plastic sleeve, then load with AN/FO and carry out blasting. Where the borehole is only moist (no free water), it would be possible to use AN/FO without recourse to plastic lining provided it is loaded and blasted immediately.

Test for Water Resistance   There are some easy-to-perform tests to establish the water resistance of AN/FO. The tests can also be done on AN prills as received from the manufacturer. Tests can be with static water or in flowing water.

In tests where water is stationary, a weighed amount of AN/FO is dropped into a measuring cylinder containing water at least four times the volume of AN/FO taken. The measuring cylinder is kept at constant temperature which is close to the borehole temperature (estimated or measured). At intervals of 1 h, the supernatant water is taken out and the AN content estimated by titration (floating FO is removed by absorbing or skimming before titration). The amount of AN leached out is calculated as a percentage and it is a measure of the leaching rate. Even the amount of FO, which has come out and is floating, can be estimated calorimetrically provided a colored dye is added to the FO before mixing it with AN. Calculation of the composition at the end of 1 h will show what performance could be expected under such real conditions of loading. The experiment can be run for different time periods consecutively.

To simulate dynamic conditions of flowing water, tests of AN/FO can be performed using a tablet disintegrator used in the pharmaceutical industry to estimate tablet dispersion in water. Small sample of AN/FO is continuously moved in a large body of water and time for complete dissolution can be measured. Speed of movement can be varied as well as temperature of the water body. This test is somewhat drastic but can give quickly an idea of approximate water resistance, especially when comparing the effectiveness of waterproofing additives. Both the above tests are comparative tests.

Effect on Performance   Due to the high solubility of AN in water even at room temperature and the fact that AN is not at all fully covered or coated uniformly with oil, which could repel for sometime at least the ingress of water, the AN gets dissolved in the water standing in the boreholes. The FO not entrenched deep in the AN prill gets leached out and separates from the AN. The composition deviates considerably from the ideal and to such an extent, total blast failure can take place. These effects are more common in bulk-loaded blasts, but where plastic sleeves or AN/FO packed in plastic cartridges sealed at both ends are used, the effect is much less severe although loading difficulties due to density of AN/FO being less than that of water can delay the loading considerably. In some cases, cartridge separation in the column charge can take place leading to misfires.

A sure sign of the effect of water on AN/FO can be observed in the color of the postblast fumes. Orange–red colored fumes, if seen postblasting even when oxygen-balanced AN/FO was loaded, shows that composition has changed because of leaching of AN in the AN/FO. The blast, even if not a complete failure, shows the effect of much lower energy generated by the explosive as compared to a dry-hole blast. Sometimes to take care of the effect of water/moisture on the explosive, additional boostering is resorted to. While this can succeed only if the effect of water leaching has only started, it does not succeed for most watery hole blasts where the explosive has been left in the borehole for more than an hour as dissolution of AN even in static water is quite rapid.

4.4.4.8 Water-Resistant AN/FO

It is logical that additives were thought of as a solution to the lack of water resistivity of AN/FO. Considerable amount of work threw up a number of additives, which could provide some relief. However due to cost considerations and effectiveness without harming the explosive performance, the choice came down to a few substances which are used in the food industry. These are natural products like guar gums of all types, carboxy methyl cellulose and starch. Synthetic oil and water-soluble chemicals such as polyacrylamides were also recommended. But it appears that commercial success has so far eluded any of these additives. Author's experience in this field brought out some important points:

1) The additive must be added to AN/FO, that is, it must be added in the end after the oil has mixed fully with AN; otherwise it prevents the oil from getting into the core of the AN prills and creates fuel imbalance.
    The additive must be rapid in building up viscosity when it comes into contact with water so that a barrier can be created between the AN/FO and water immediately.
2) A cross-linker could be used for gums that links up the guar molecules. The cross-linker has to act slowly only after the guar gum has attained full hydration and hence using a cross-linker can be useful if the AN/FO stays in the watery hole for longer periods.
3) A polymer if used must be oil soluble and capable of being cross-linked to provide a rigid matrix which can act as an effective barrier. While the polymer should develop viscosity rapidly, cross-linking can proceed continuously and slowly. Too fast a cross working will produce a rigid matrix of AN/FO and may affect the loading characteristic.
4) In bulk loading, provision in the truck has to be made for addition of specialty chemicals before dropping into borehole.
5) Additives can be hygroscopic by themselves and for most natural products to be effective need to be kept in dry condition before addition.
6) Additives should not interfere with the free flow and loadability of the AN/FO.

The fact that a waterproof-resistant barrier created on particles of AN/FO is a good means of solving the problem of its low water resistance was responsible for the development and commercial use of emulsion/slurry explosive itself as a coating substance. Such products known as heavy ammonium nitrate/fuel oil (HANFO) are a commercial success in the bulk-loading blasting applications.

4.4.4.9 Increasing the Energy of AN/FO and its Fume Characteristics

Fume characteristic of an explosive is measured in terms of the volume of toxic gases produced when it undergoes detonation. This differs from explosive to

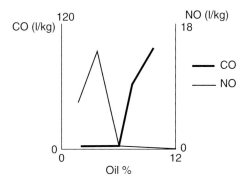

Figure 4.10 Toxic gas volume versus oil content in AN/FO.

explosive and also for the same explosive under different conditions of blasting. When the explosive is fully oxygen balanced and detonated ideally, least amount of toxic gases are produced (Figure 4.10).

Toxic gases are NO and CO. Although CO_2 is not included under toxic gases for purposes of determining the fume category of the explosive, in practice high volumes of CO_2 are detrimental to human life. Prolonged breathing or heavy dose of CO_2 can incapacitate a human being due to suffocation (lack of oxygen). The presence of CO_2 even after fumes are cleared could persist in dips and hollows due to its High Density (1.5) as compared to air.

Fume categorization of explosives are done according to the amount of toxic gases produced when a known quality of the explosive is set off under confinement in a chamber totally sealed and capable of withstanding the heat and pressure generated during its detonation. Such an apparatus accepted as benchmark is the Bichel gauge; ratings are from A to C depending on the volume of toxic gases produced. Commercial literature gives fume category in descriptive terminology such as good, very good, or excellent.

AN/FO has been subjected to extensive studies regarding generation of toxic fumes when it explodes in large boreholes in open cast mines and in underground mines in small diameters. The studies showed that even though fully oxygen-balanced AN/FO was used, postblast fumes contained toxic gases several degrees higher in concentration than expected. It is now known that deviation from ideal detonation due to various factors lead to enhanced toxic fumes production.

These are lower charge diameter, inadequate water resistance, inadequate priming, premature loss of confinement due to stemming blowing off or cracks developing in the borehole, AN prills conditions, nonuniform mixing of AN/FO, and High Density (nearer to critical density) leading to lack of sufficient number of hotspots.

It is not possible in practice to ensure that all the factors stated above are under control, and hence in practice it is assumed that a certain quantity of toxic gases are always produced and the personnel are made to wait for the clearing of the postblast fumes till they are allowed near the mine face. This is enforced in the

Table 4.7 Contribution of AL to strength of AN/FO.

% AL[a]	Bulk density of explosive	Weight strength	Bulk strength
0	0.83	100	100
2.5	0.85	110	110
5.0	0.86	118	120
10.0	0.88	133	138
12.5	0.89	139	147
15.0	0.90	146	155

[a]AL = aluminum.

underground mining strictly, even in mines with good ventilation, and can delay the period between two blasts affecting productivity.

Ideally detonation of AN/FO will produce 50 l/kg at $0\,^\circ$C at density of AN 0.85 g/cc with hardly any trace of CO and oxides of N_2. However nonideal detonation can produce NO, CO, NH_3, NH_4NO_3 aerosol in addition to N_2, H_2O, and CO_2 leading to low energy output.

This is due to under- or overfueling and surface decomposition of AN. While AN/FO provides the cheapest available energy per unit weight of explosive, it may be desirable for some conditions of blasting to generate more strength at some extra cost. This is best achieved by addition of reactive metals as fuel and allowing it to get oxidized in an exothermic reaction. Aluminum has established itself as the most optimum additive with gains in energy and density as seen in Table 4.7.

The effect of aluminum is not linear (Figure 4.11) and tapers off at the higher percentages (beyond 10%). If cost/benefit is also taken into consideration, addition beyond 6% of aluminum appears to bring little extra in terms of performance. Theoretically the reaction

$3NH_4NO_3 + 2Al = 3N_2 + 6H_2O + Al_2O_3$(solid) is expected to release 6800 kJ/kg

of heat energy as compared to 3900 kJ/kg without aluminum.

In practice, the total required fuel component is partly supplied by FO and partly by aluminum.

Theoretically 1% aluminum addition brings in 4.5% increase in the energy of the explosive. In practice, it is only 2.5% due to side and competing reactions, after burning rather than reaction with NH_4NO_3. The physical form of the aluminum added is also very important to get the best energy output possible. The finer the particle size and larger the surface area, the more reactive the aluminum, but at the same time these finer powders exhibit greater reactivity to moisture and dust and fire hazards unless protected. The finer powders are also costlier. Hence today the best compromise is to use fine atomized powder or finely chopped aluminum foil. More details on aluminum usage in explosives is given in a later chapter of this book.

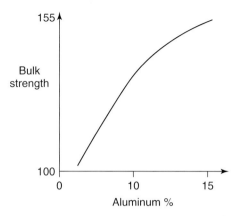

Figure 4.11 Effect of aluminum in AN/FO.

The use of molasses along with aluminum was quite popular for producing flowing AN/fuel explosive and the sugar in molasses acted as a good fuel. Since molasses was liquid, it could mix with AN even though not as well as diesel oil because of its higher viscosity. Question of stability also became an issue for such explosives since molasses autodegrades and is catalyzed by aluminum. Also aluminum/H_2O reaction prematurely due to the H_2O content of molasses caused hazards due to runaway reactions during manufacture and instability problems in storage. However, the explosive had good energy output and initial density was higher than AN/FO. Current situation of not so easy availability of molasses due to it being a good source of alcohol has more or less seen the end of molasses-based, aluminized AN explosives (molanal).

4.4.5
Safety Considerations in AN/FO

Although AN/FO is relatively insensitive to impact and friction established by standard fall hammer and torpedo friction tests, static electricity discharged directly onto the AN/FO has indicated possible hazard situation. The hazard arises due to fire developing on the explosive due to diesel oil fumes present on the surface of the explosive, leading to continuous burning and explosion if the AN/FO is in a confined situation such as a storage bin or magazine. Diesel fuel, even though flash point is greater than 60 °C, catches fire quickly and hence has to be kept away from AN and AN/FO in underground storage.

AN/FO of course is sensitive to shock from another explosive detonation wave. Thus observing safety distances becomes mandatory. AN itself, even though classified as a nonexplosive, is known to ignite and explode on receiving sufficiently high impact by hot missiles and shock wave pressure from a nearby explosion.

Complacency based on the so-called nonexplosive classification of AN and low hazard risk associated with AN/FO has led to somewhat casual handling of AN/FO manufacture resulting in devastating accidents.

4.4.6
Summary – AN/FO Explosives

It is seen that AN/FO being a simple two-component explosive can function as a good explosive in medium/large-diameter holes due to its ease of manufacture and loading. AN/FO possesses adequate gas volume and energy if detonating close to ideal. No doubt AN/FO suffers from poor water resistance, bad fume characteristics, and low detonation pressure, but in spite of these disadvantages, AN/FO is the most used AN-based explosives volume wise globally. The performance of AN/FO is heavily dependent on the type and quality of AN used as raw material and uniform mixing with FO. Special types of AN when used in AN/FO can make the AN/FO detonate more efficiently in smaller diameters. Performance of AN/FO is dependent on many extraneous factors in blasting, but by and large it is still the most cost-effective product. When used underground, inadequate fume characteristics and low density become an issue.

Very low density AN/FO is made by using AN prills in which microballoons are entrained in individual spheres. It is claimed that such an AN/FO has better performance properties than standard AN/FO. Advantage is claimed in such a product having higher detonation pressure (VOD), full release of energy, less toxic fumes, and improved sensitivity to initiation due to the incorporation of the hollow microspheres, which can function effectively as additional hotspots.

Low-energy and low-weight-strength AN/FO is made by addition of expanded polystyrene beads in the composition. This product is useful in smooth blasting operations.

Aluminum used in AN/FO for enhanced energy is usually atomized powders or granules. While aluminum flake powder can also function as energy enhancer, its cost is not commensurate with the increase in energy obtained. Further handling of aluminum flake powder in the dry form brings additional fire dust explosion hazards. Aluminum chopped foil can also be used and in USA scrap from the aircraft industry was an energy source. Technically the energetic metal has to react rapidly and not to show any induction period to ignite once the detonation has started and heat has been released from explosives decomposition; otherwise the energy of the metal is wasted in afterburning, that is, the metal combines with oxygen of the atmosphere after the explosive is consumed in the detonation process and does not contribute to the enhancement of the explosive's energy. The aluminum combines with oxygen forming Al_2O_3, which is a sold residue and reduces the gas volume as compared to nonaluminized AN/FO, but the heat energy released which is five times greater compensates the loss in gas volume. Finally it is all a question of finding the cheapest source of useful energy-giving metal.

4.4.7
Quality Checks

The checks required to maintain consistently good-quality AN/FO during manu-facture are not many but vital. AN being the major and most critical raw material needs to be checked before use. The most important properties to be checked are

1) prill size and distribution,
2) porosity (oil absorption), and
3) phase stabilization.

A quick Differential Scanning Calorimetry thermogram to establish the temperature of IV to III transition gives an idea whether adequate stabilization has been obtained or not (transition temperature should be above $50\,^{\circ}$C and not $32\,^{\circ}$C). In Fuel Oil (FO), flash point composition as given from reputable manufacturer can be accepted. At the most viscosity can be checked.

During manufacture, the most important criterion is to maintain the correct proportion of AN and FO. If a continuous process is used, then flow of both needs to be regulated to keep within the set limits. Generally the oil content is kept within 6.0–6.5%. The addition should be in the form of fine spray for better distribution. In batch process, the quantities of AN and FO used per batch should reflect 94/6. Mixing time should be constant and uniformity is checked by taking samples and comparing calorimetrically (dye should be used in oil). Once the mixing time is established, it should be kept the same as long as the same source of AN continues. If changed, standardization would be needed again. All additions of other ingredients are to be added after AN/FO has been formed, that is, after FO addition and mixing is complete.

On the finished product (AN/FO), the quality checks are pourability, oil content, and water resistance. Performance tests for release of product for blasting for batch process and cartridged product are measuring VOD in a diameter less than the borehole diameter where blasting is to be carried out. For instance if the boreholes used are 5 in. diameter, it is best to test in 3 in. diameter thus giving adequate margin for unknown adverse factors. Primer quality and quantity for all VOD measurement and release tests should be standardized and not varied. For bulk-loaded AN/FO, performance test for approving product for use needs to be done prior to actual loading into the borehole. The speeds at which bulk loading is done exclude online checking. Once in a while if underwater testing facilities are available, the AN/FO could be checked for total energy and its components. This is particularly useful and necessary if AN/FO contains metal powder additives to establish cost benefit in terms of enhanced energy output obtained by addition of the costly energizer. Observation of the blast itself is a very good feedback on the performance of the AN/FO being manufactured, although there are many extraneous factors not related to AN/FO that influence the performance. They are related to drilling pattern, rock strata, availability of free face, delay sequence, stemming, and presence of water. Related directly to AN/FO would be its loading density (beyond 1.15 g/cc even in large diameters performance drops off significantly), its water resistance if used in moist/watery holes with external protection (plastic sheath), and its oxygen balance. If postblast fumes show copious orange red colored fumes – not to be confused with particulate material cloud which could sometime be also red in color – shows excessive positive oxygen

Table 4.8 Energy loss in AN/FO based on oxygen balance.

Oxygen balance	Oil content (%)	Energy loss (%)	Effect on blasting
Ideal	5.7	None	Best results
Low fuel	5.0	5.3	Orange fumes
	4.0	12.1	Bad performance
	3.0	20.0	Unacceptable
High fuel	7.0	1.5	Dark gray fumes
	8.0	2.9	—
	9.0	4.9	Low yield

balance. Gray fumes show excessive negative oxygen balance and loss of energy in either case (Table 4.8). Especially underfueling will reflect in the blast output being unsatisfactory. In some instances factors mentioned earlier could lead to nonideal detonation even in a perfectly oxygen-balanced AN/FO resulting in excessive toxic fumes generation and low-energy output as shown in Figures 4.5, 4.6, and 4.10.

Hence it is seen that not only the correct ratio between AN and FO needs to be used in the composition but also the mixing has to be efficient to ensure that the FO is evenly distributed over the AN prills. As mentioned earlier since loss in energy is much less when the fuel is in excess than when it is below the ideal value, it is always prudent to provide slight excess of fuel in the AN/FO.

References

1. Oommen, C. and Jain, S.R. (1999) *J. Hazard. Mater.*, **A67**, 253–281.
2. Urbanski, T. (1983) *Chemistry and Technology of Explosives*, Pergamon, Oxford, pp. 450–475.
3. (1976) *Comprehensive Introduction of AN/FO*, Mitsubishi Chemical Industries Limited, Tokyo, Japan.

Mishra, I.B. (1986) Phase stabilization of AN with KF. US Patent 4552736.
Konya, C.J. (1995) *Blast Design*, Intercontinental Development Corporation, Montville, OH
Porter, P.H. (1984) AN blasting agents, quality testing for maximum benefit, SEE. 10th Annual Conference, Florida.
Rowland, J.H. III and Mainiero, R. (2000) *Int. SEE*, **I**, 163.

Further Reading

Cook, M.A. and Taylor, A. (1951) *Ind. Eng. Chem.*, **43**, 1098.

5
Slurries and Water Gels

5.1
Development

In the mid-1950s as ammonium nitrate/fuel oil (AN/FO) was gaining ground at the cost of dynamite, especially for large-diameter blasting, some limitations regarding availability of the desired quality of prills and some performance characteristics such as poor water resistance and low velocity of detonation (VOD) were causing concern. It was a remarkable out of the box thinking from Prof. M.A. Cook in USA (1958) that led to the discovery that water which was shunned as a damaging component could be used advantageously and slurries were born [1].

Slurries are supersaturated oxidizer solution and fuel mixtures with AN and H_2O as major components. Thickeners are used to get the desired consistency. They also have additives to increase the energy. The basic concept of making a good explosive wherein the oxidizer and the fuel are brought into intimate contact and inclusion of hotspots in the form of microbubbles was kept in mind while designing a viable slurry. Later on to cater to the needs of packaged product and small-diameter, cap-sensitive explosives, a more rigid product as compared to the flowing slurries was designed and established. This was known as *water gel*. Initially both slurries and water gels did contain self-explosive ingredients like trinitrotoluene (TNT), but continuous attempts were made, in particular in large-diameter products, to eliminate the use of self-explosives, thereby restoring parity as far as transport classification was concerned with dry blasting agents like AN/FO. Considerable research in establishing of new ingredients, new formulas, and processing [2, 3] resulted in the production of large quantities of slurries and water gels. United states bureau of mines also did an enormous amount of pioneering work in evolving cap-sensitive products without self-explosive ingredients to compete and replace dynamites for use in coal mines [4–6].

5.2
Design

Based on the principle of bringing together intimately oxidizer and fuel and sensitizing them, slurries were formulated. Water was used as a medium for

Ammonium Nitrate Explosives for Civil Applications: Slurries, Emulsions and Ammonium Nitrate Fuel Oils,
First Edition. E.G. Mahadevan.
© 2013 Wiley-VCH Verlag GmbH & Co. KGaA. Published 2013 by Wiley-VCH Verlag GmbH & Co. KGaA.

achieving the above and the oxidizer consisted mostly AN. However the AN was not required to be in prilled form since the AN was dissolved in water during the process and reappeared in crystalline form whose size and structure depended on the presence of other ingredients also. A great advantage was found in that these products could be tailor-made for energy and density with a range higher than that of AN/FO. Also the manufacturer had a wide choice of the ingredients which could be used giving a lot of lee way for controlling the cost. One could switch the ingredients in the same functional category in order to overcome scarcity situations.

5.2.1
Large-Diameter Packaged Product (Water Gels)

These are for blasting in above 50 mm diameters and sensitivity to No. 6 detonator is not a precondition. Initiation is achieved by boostering with high explosives like pentolite or with a cap-sensitive product. Without the presence of self-explosive ingredients, these are classified as blasting agents and can be easily transported. Water gels are slurries in rigid/solid or semisolid (nonflowy) physical state and are preferred for packaging and transporting as they will not flow out even if the package bursts open. Also water gels have a much longer shelf life.

A product made out of oxidizer consisting of mostly AN with $NaNO_3$ or $Ca(NO_3)_2$ and fueled by sugar or ethylene glycol, sensitized by air bubbles through chemical gassing, thickened by gums initially, and cross-linked duly is one of the cheapest slurry products capable of competing with AN/FO on cost and strength basis.

Stronger products have energy-giving additives like aluminum (AL) granules, Fe powders, silicon, AN prills, sulfur, and per chlorates in the composition.

5.2.2
List of Ingredients

List of ingredients commonly used is given in Table 5.1. The list is extensive but by no means comprehensive as there are many more being added continuously.

5.2.3
Small-Diameter, Cap-Sensitive Water Gels

These need to be rigid products with long shelf life and hence only fully cross-linked water gels satisfy this requirement. Cross-linking partly has to take place during the process and continues in the package for a few days. Since the product needs to detonate efficiently on initiation from a No. 6 detonator, there should be adequate sensitization achieved either through incorporation of air bubbles mechanically while processing or through chemical gassing or by addition of metal sensitizers with low bulk density like flake AL powder. Organic sensitizer such as monomethylaminenitrate is also used by some to provide not only cap and propagation sensitivity but also to contribute to the strength of the explosive.

Table 5.1 Ingredients commonly used in AN-based slurries and water gels.

Oxidizers	Fuels	Sensitizers
Ammonium nitrate	Aluminum	Methyl amine nitrates
Ammonium perchlorate	Ammonium oxalate	Ethylene diamine dinitrate
Barium nitrate	Ammonium acetate	Aluminum flake powder
Calcium nitrate	Carbon	Microballoons
Calcium perchlorates	Calcium stearate	Air bubbles
Magnesium nitrate	Diethylene glycol	Sulfur
Nitric acid	Mono ethylene glycol	AN porous prills
Potassium perchlorate	Glycerol	Nitro methane
Sodium perchlorate	Urea	
Sodium nitrate	Sugar	
	Sulfur	
	Hexamine	

These products are replacements for dynamite in underground blasting in coal mines and non-coal-mining of all kinds requiring controlled blasting. They are also used in the open for quarrying and well digging.

Where permissible products are needed coolants are added to keep the flame temperature down without impairing significantly the explosive strength.

5.2.4
Bulk Delivery Product

These are made flowy as they have to be pumped and hence these could be classified as slurries. The degree of cross-linking of the thickener (gum) if at all done is slight, just sufficient to prevent segregation of the solids from the body of the explosive. Bulk slurries are produced both in stationary plants and pumped directly into boreholes by using pump trucks. Nowadays nonexplosive slurry matrix is produced in the stationary plant and loaded onto bulk delivery vehicles, which convert the nonexplosive matrix into explosive product by addition of sensitizers before discharge into the borehole. Additional cross-linkers could also be injected at this stage for increased waterproofness.

5.2.5
Basic Concepts of Formulation

5.2.5.1 Oxygen Balance
Just like in AN/FO zero oxygen balance (OB) is the ultimate objective. The final OB of the explosive is the sum of the OB of all the ingredients.

Even thickeners like guar gum (GG) are considered while calculating the OB. Some typical basic product formulations (nonpermissible type) are given in Table 5.2.

Table 5.2 Compositions of different types of slurries/water gel explosives.

Ingredient (%)	Nonaluminized	Aluminized	Cap-sensitive	Bulk
Ammonium nitrate	57.5	62.5	57.5	67
Sodium nitrate	10.0	8.0	10.0	5.0
Water	15–18	15	12–14	16
Gums	1.5	1.2	1.2	0.8
Sugar	13.0	8.0	—	—
Ethylene glycol	—	—	6.0	2.0
Urea	—	2.0	4.0	6.0
Aluminum	—	6.0	4.0	1.0
Sulfur	—	—	3.0	2.0
Oxygen balance	0.0	0.0	−3.0	0.0

The range of percentages indicated above leaves considerable flexibility for formulating a product with the required density, energy (weight strength), viscosity, slurry or water gel, stability in storage, VOD, and gap sensitivity.

5.2.5.2 Thumb Rules for Design

Some basic thumb rules in formulating are given below:

1) As oxidizer using straight AN alone is not recommended as sensitivity to initiation, speed of chemical reaction, storage stability are not favorable. If at all necessary for some reason to use only AN as oxidizer, it is used in bulk slurry only where the product is blasted after loading within a short time and the blast is done in very large diameters (not less than 175 mm diameter) with substantial boostering.
2) Always use a combination of oxidizer salts with AN as major component. Such mixtures contain $NaNO_3$, $Ca(NO_3)_2$, and $NaClO_4$ as other oxidizers.

Sodium nitrate has the property of fast chemical reactivity and it helps in lowering the critical diameter, promoting a higher VOD, lowering the initiation requirement.

$Ca(NO_3)_2$ has been shown to be somewhat slower than either AN or sodium nitrate in its decomposition chemistry but has the advantage of keeping AN in solution by forming eutectic mixture where crystallization point is depressed due to increased solubility. In such cases where calcium nitrate is used, the processing temperatures can be lowered to correspond with crystallization point of the OB. Use of CN has been proven to increase the cold temperature sensitivity of water gels/slurries again due to property of CN of keeping the AN in solution at lower temperature than when CN is not present. Use of CN alone or as a major component is not recommended as strength falls off due to its lower enthalpy and heat of formation. However use of CN lowers the explosion temperature and hence many formulas for permissible explosives do contain CN. Calcium nitrate interferes with the hydration of GGs and when CN is used in larger percentages (beyond 3%), it is necessary to use modified gums for attaining the required viscosity while processing and for stable gel later on. It is also cheap and in many explosive units where dilute HNO_3 is available after nitration, calcium nitrate is produced in-house by chemical reaction with limestone.

Use of perchlorates can bring in additional sensitivity and oxygen availability. They are also safer to handle than chlorates especially in aqueous solutions. Sodium perchlorate in combination with organic sensitizer such as MMAN is claimed to give additional benefits in speed of reaction due to formation of methylamine perchlorate and its subsequent thermal decomposition. Use of perchlorate does bring in some thoughts about safety since the end product can show increased sensitivity to impact, friction, and heat especially in the dry state and at very low H_2O contents in the slurry/water gel. Since the water in water gel is not easily driven off, perchlorates in solution can be handled on a routine basis with no great risk. The use of sodium perchlorate is done firstly for producing cap-sensitive product used in hard rock and when explosive needs to remain cap-sensitive for a longer period. The presence of chlorine in the perchlorate could function as a coolant and flame suppressor and permissible explosives formulations can contain perchlorate with some benefit. Sodium perchlorate is an industrial chemical and not very costly and the cost/benefit in the explosive formulation is very acceptable.

The presence of CN, SN, and sodium perchlorate in OB raises OB density substantially and lowers the crystallization point.

Density of AN/SN/H_2O	1.40 g/cc	Crystallization point	59 °C
Density of AN/SN/CN/SPC/H_2O	1.56 g/cc	Crystallization point	44 °C

5.2.5.3 Role of Water

Use of water is normally associated with deleterious effect in explosives. However in slurry/water gel compositions on one of the most unusual discoveries water has helped formulations in many ways. The role of water in slurries, water gels, and emulsions is manyfold. Firstly water is the medium in which the oxidizer salts, fuel, and sensitizers come in close contact with each other.

Secondly water adds to the strength in that it contributes to the gas volume by converting itself into steam. Further water is an inexpensive component and helps in regulating the rheology of the explosive in conjunction with thickeners. It serves as a matrix in which hotspots are embedded, be they air bubbles or microspheres. Water also serves as a coolant and is a necessary component of permissible water gels/emulsions. In other words, water is a valuable and an indispensable component, as far as water-based explosives are concerned.

Going into details, let us examine how water acts as a facilitator in an AN explosive. Solubility data shows that most inorganic oxidizers are highly soluble in water and they have a positive temperature coefficient, that is, solubility increases with temperature (except for KNO_3) (Table 5.3).

Table 5.3 Solubility of oxidizer salts in water at different temperatures.

Temperature (°C)	Solubility (g/cc) of ammonium nitrate	Solubility (g/cc) of sodium nitrate	Solubility (g/cc) of sodium perchlorate
0	20	70	60
10	150	80	—
20	190	90	—
30	230	—	70
40	280	100	—
60	400	125	74
80	620	—	—

AN particularly shows a steep increase in solubility with increase in temperature especially at the lower range (0–30 °C). It can be further improved by developing eutectics *in situ* by addition of other salts like $Ca(NO_3)_2$, SN, SPC, and urea. Fuels, which are also soluble or miscible with water-like formamide, can be used to develop a hetero/homogeneous phase where part of the oxidizer salt is in the solid form and partly in solution in contact with fuel and sensitizer. This is achieved by taking all the AN into solution at a particular water content in the composition by heating to a temperature above the crystallization point and then cooling the solution. AN will crystallize out; the quantity and type of crystal structure will depend upon the ambient temperature. It is important that AN crystallizing out is sensitive and able to function as an explosive. This is achieved if the AN coming out is in the form of small crystals but which are discrete; in other word they are not bridged to form a cake/crust. Use of AN crystal habit modifiers, phase stabilizers, correct mixing parameters, and providing a physical barrier in the form of a viscous and strong film between the emerging crystals will ensure that the AN crystallizing out is and remains in a state conducive to steady-state detonation.

One of the easiest ways would be to keep maximum amount of the salts in solution by increasing the water content. If this is to be attempted at ambient temperature say 25 °C, so much H_2O would be required that the explosive would be extremely weak and would not be of use in any applications except perhaps in the highest category of permissible explosive. On the other hand for competing with powerful dynamites in terms of weight strength, density of AN, and other oxidizer salts content will have to go up at the cost of water content.

Role of water is also critical in processing. Ease of processing and safety factors indicate that there is a trade-off. All other factors being same, a composition with lower water content will need higher processing temperature and even the OB will have to be kept at the higher temperature. This means more steam and energy consumption adding to costs. Low water compositions are also more sensitive to impact and friction. Role of water in lessening safety hazards normally encountered in pumping cannot be underestimated.

Role of water is also very critical in controlling the viscosity of slurry for bulk delivery systems and pumping into boreholes.

5.2.5.4 Basic Composition and Process

General compositions followed today in industry based on the pioneering work done in 1960s is to achieve an explosive with maximum available energy at the highest density and lowest critical diameter and in a physical condition suitable for packaging/and for bulk delivery.

The process by and large is based on thickening a hot oxidizer solution by means of a thickener like GG, adjusting the density through mechanical aeration or gassing, mixing the solid ingredients for enhancing the energy to arrive at the finished product by cooling. The proportions of the various ingredients are adjusted to satisfy field requirements.

As of today efforts are still continuing to reach the benchmark of dynamites ($d = 1.45$) without reaching the hazard sensitivity of compositions containing nitroglycerine (NG).

Basic process is shown in Figure 5.1.

Compositions can usually fall within certain ranges depending on the product required given in Table 5.4.

Thus it is seen there is a wide range of compositions possible, but always the final composition made into explosive is close to zero OB or slightly negative. In packaged slurry, the question of whether to take all the polythene/paper into

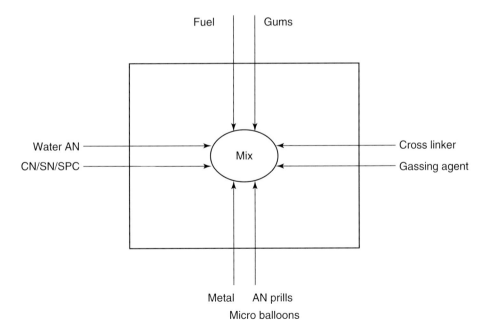

Figure 5.1 Basic process diagram.

Table 5.4 Range of ingredient quantity in water gel explosives.

Ingredients	Quantity (%)
Oxidizer salts	45–85
Water	8–25
Thickeners	1.0–2.5
Cross-linkers	0.05–0.10
Fuels	6–15
Metals	3–8
Organic sensitizers	10–30
Gassing agents	0.15–0.5
Solid AN prills	5–15

reaction and add to the fuel value or consider it only for after burning has been debated and some studies performed, but no final conclusion could be arrived at. It is therefore safest to exclude the influence of packaging material in the OB calculations. In any event except for the smallest cartridge size, it hardly makes a difference. In a 150- or 200-g cartridge, paper/polythene weight could be 4–6 g, which amounts to 3% of explosive weight, a significant amount. In case of larger diameter products (say 50 mm diameter and 300 mm length) where 1 kg of explosive is packaged, this comes down to 1%, in 75 mm diameter (2 kg) it is 0.6%, in 125 mm diameter (6.25 kg) 0.4%, and in 175 mm diameter (12.5 kg) 0.2% only. Formulating an oxygen-positive composition and expecting fuel contribution from the packaging material to make an overall balanced composition is very chancy and unwise as practical experience has shown. In most test firing of such a product, there were remnants of packaging material left behind after a blast and the explosive performed poorly.

5.3
Process Technology

5.3.1
Batch Process

This was the first process adopted for large-scale production of water gels and in stationary plants for production of bulk loading slurries.

Heart of the process is the batch mixer. Oxidizer is made in large tanks with slow-speed stirring and recirculation facility. All tanks and lines are insulated and jacketed. Temperatures maintained can be up to 90 °C. Motors are flame- and explosion-proof. The plant will be compact if it can be multilevel. The mixer is placed at the top or mid level. Solid ingredients can be added through a screen in weighed amounts from the top into the mixer. For drawing in the required amount of

Table 5.5   Blender occupation time (batch vs. pre hydration).

Operation	Standard batch time taken in minutes	Prehydrated batch time taken in minutes
Oxidizer blend feed	6	4
Hydration in blender	12–15	6–8
Aeration (mechanical)	5	3
Solids addition and mixing	10	10
Blender occupation	35	25

OB, volumetric vessel can be used. Preweighed and mixed thickeners/cross-linkers in bags can be added directly into mixer as also other solid ingredients in small quantities. Aluminum powder or paste preweighed can be added in a closed system. Batch size is 500–600 kg.

Packing equipment is arranged at the ground level so that packed explosives can be removed easily and continuously or pumped into bulk delivery vehicle.

The basic batch-mixing process implemented in the 1970s works even today effectively, but manpower required to be deployed is high for all the operations. A higher degree of sophistication involving higher capital outlay by introducing automation like flow meters, blend tanks mounted on automatic scales and vibratory feed facilitators for solids addition, and automatic packing requires much less manpower and is practiced in countries where manpower is costly. Contact materials are made of standard stainless steel (SS 316). The blender is usually a double helix type with central discharge, capable of forward/reverse mixing at variable speeds from 50 to 600 rpm.

An improvement over the basic batch process consists in introducing a prehydration vessel, which separately operates and provides a partially hydrated oxidizer solution to be added to the main mixer containing the rest of the OB. This speeds up the batch-making time frame and also provides better mixing of GG (thickener). No cross-linker is added in the prehydration vessel. Table 5.5 shows a comparison of batch time for the two processes.

Thus the main blender can be kept continuously occupied by providing prehydrated material from a separate vessel since time for prehydration is much less than batch-making time in the main mixer. Prehydration also ensures good and complete hydration as the thickener does not come into contact with cross-linker or other solids of previous batches till it is well hydrated. This reduces aeration time. Depending on the final product (gel or slurry) prehydration time, quantity of thickener can be controlled easily in the prehydration vessel, which is nothing but a jacketed fixed-volume vessel with a variable speed, vertical dispersing stirrer. OB (fixed quantity) is pumped in and thickener added and dispersed, hydrated, and then fed into the batch blender and mixed with rest of the OB and duly processed with a view to speed up the batch-making process and increase the throughput in the blender for increased production. A 30% increase in productivity can be achieved by using prehydration process.

5.3.2
Continuous Process

Continuous processes have been developed wherein the hold up of explosive material is considerably reduced. In fact the development of continuous process (by IRECO, USA 1978) led to the introduction of sophisticated bulk delivery trucks where the slurry explosive could be manufactured on the mine face itself and delivered instantaneously into the borehole.

The principle of most continuous processes is based on accurately bringing into small high-speed mixer quantities of oxidizer, fuel, and other additives in the right proportion as determined by the formula. The mixing time is just 1 min. The raw materials (RMs) are fed to the bottom of the continuous mixer, gets mixed, rises to the top, and is pushed out. Various proprietary designs of mixers have been invented, fabricated, and used.

The continuous mixing can also be used in a stationary plant instead of a batch-mixing system. Advantages are the same as before – less explosive hold up and faster processing. However all feed systems have to be accurate, otherwise imbalances can occur. Continuous monitoring of flow rates with alarms is necessary. Quality of mechanical/chemical aeration as per author's experience is not as good in the continuous process as in the batch process. The air bubble size and quality of the gel with respect to stability are also not ideal probably due to the fact that hydration of the gum in that short period of residence in the continuous mixer is not complete. Hence it may be prudent to use this high-output continuous process for making lower density (LD) booster-sensitive products only.

Semicontinuous process employs aeration, while the matrix at high density made in a batch process is moved through a pipe with air injectors and an on-line static mixer. This process is good to use along with gassing agent for LD applications both for packaged and bulk slurries.

5.3.3
Packaging Systems

For producing cartridged product, packaging of slurry/water gel is done in polythene film, polythene rigid tubes, or paper tubes. The basic process here is to move the material stored in the hopper and push it out into the empty cartridge shell. If the material is flowy gravity feed alone is sufficient, but in case of advanced cross-linked product it is necessary to employ a roto pump for pushing the material into small diameter cartridge shells in paper or rigid poly tubes. The system is same as used for packing LD explosives except that nozzles are more and smaller in diameter. The nozzles can be 16–20 in numbers and are arranged either in a straight line or in a circle or in a hexagon like the Du'Pont machine used for packing dynamite.

The empty tubes are arranged in a wooden or AL frame. The entire frame is kept below the small-diameter packing machine in line with the discharge nozzles. The frame is then lifted up so that the nozzles go into the bottom of the tubes at which point individual piston or air pushes the material into the bottom of the tube. As the tube is filled from the bottom, the tube-holding frame is slowly withdrawn downward till the material fills the tube to the brim and without any voids. The stroke is adjusted to fill the volume of the individual tubes. Usually 125 g in 25-mm-diameter tubes, 150–200 g in 32-mm-diameter tubes are packed depending on the density of the product. This type of machine (designed in house) is called *tray packer*. The lifting of the tray/frame can be done manually or assisted by motor. The height of travel of the tray can also be adjusted. After the tubes are filled, they move to the crimping and capping station where they are crimped if they are paper tubes or press fit caps put on if they are rigid poly tubes. Again these operations can be done manually or through machines. After closure, the cartridges are wiped clean and packed in individual cartons of 2.5 kg each. Nine to 10 such cartons are assembled in a shipping case. Use of wooden cases has been given up and only cardboard cases are used. The number of individual cartons and cartridges per unit case and the total weight per case are according to regulations existing locally for local dispatches and according to international transport of dangerous goods regulations for export. For large-diameter packing, strong polythene films clipped at one end are filled by means of a hydraulically assisted piston packer and after getting the correct weight into it, the other end is also clipped using poly-coated wire.

A type of form fill machine used for sausage filling can also be used for packing explosives. These are known as *chub style packing* and the finished cartridge has clips at both ends. The machine operates continuously as long as the explosives feed is continuos. The special polymer film is drawn from a roll, sealed into a tube of the required diameter, clipped at one end, required quantity of material put into it, and the other end is also clipped. The cartridges are coming out continuously jointed and they are separated into individual cartridges by a cutting knife.

These machines are sophisticated and automatic with controls for flow, number of cartridges per minute (speed), for sealing operation, and so on. The speed of packing is very high. While manual tray system may produce at best 10–12 kg/min, that is, about 600 kg/h, the chub machine can pack 1.5 tons/h (1500 kg/h) – a factor of 2.5 times more with a substantial reduction in manpower. Some machines claim even higher outputs of up to 300 individual cartridges per minute (2.5 tons of explosive per hour). The finished cartridge also looks excellent when made in the form fill machine, but the cartridge is as soft or hard as the material inside. The same machine can be used for different diameters within a range of 25–75 mm.

Important points for consideration to achieve safe and clean packing are as follows:

1) The product made for packing must be of the right viscosity and capable of flowing down if gravity feed is used or capable of being pumped if pumps are used.

2) As the product viscosity invariably due to progress of cross-linking and lowering of temperature in the hopper (unless hopper is kept at a constant temperature) will change, the criteria in (1) has to be fulfilled till the material from a particular batch is packed out fully when batch process is employed. This would have to be taken care of by adjustment of flow at the packing end as otherwise cartridges of varying weight/length can result. Change in density continuously after production, if chemical gassing is used, also contributes to the variability of the product density and adjustments need to be made so that constant weight of product is delivered into the individual cartridge.

3) In the form fill machine, the sealing of poly film is by hot air or by a sealing adhesive. For both, it is necessary to see that sealing is perfect; otherwise leakage of the material will occur and spoil the entire package. Clipping also should be tight and leak proof.

4) Care must be taken to see that lines are free and not choked when pumping is going on; otherwise dangerous pressures can be built up in the line and could result in a hazardous situation apart from no feed to the packing machine resulting in half-filled or empty cartridges.

5) The clipping mechanism needs to be kept free of explosive material at all times during operation. Since there is considerable pressure exerted during the clipping process, any explosive material if present on the clipping mechanism will be subjected to intense pressure and also come in between metal parts again leading to a hazards situation.

6) There is a possibility of static electricity being produced and discharged during the movement of the film on the tube forming mandrel. Earthing of the machine will render it safe as the generated static electricity will be conducted away safety.

7) If the material is becoming cap-sensitive explosive by reduction in density through chemical gassing in the packing machine, then there is a need to consider erecting a barrier or separation of the packing machine from the hopper by distance as it also introduce discontinuity in the line conveying the material into the packer so that any explosion in the packer is not transmitted back into the hopper containing a large amount of explosive.

8) Various types of films made out of different polymers are offered each with different advantages. The right type of film for the right product needs to be selected, the most important being impermeability (air or gas should neither diffuse out nor come in) and inertness in contact with the explosive ingredient used. No stretching and loss of strength of the film should occur on storage; otherwise this will lead to limp and out-of-shape cartridges causing difficulties in priming and loading.

9) Operator should obey all safety precautions given in the manual of the packing machine. In brief, he should wear goggles, gloves, conducting shoes, cotton overalls, and follow the operating manual.

5.4
Quality Checks

5.4.1
Raw Materials

All RMs need to be checked against specification. AN prills can also be used to make the oxidizer solution, but generally it will be more expensive than other types of AN like lumps, solids, or crystalline. Most units use AN melt (80% AN 20% H_2O or 93% AN 7% H_2O) brought in insulated tanker and unloaded into insulated storage tanks from which measured or weighed quantities are drawn into the oxidizer blend making vessel. In case of solids, the AN should not contain any nitrites as this can lead to premature gassing and possibly a runaway reaction with AN. Moisture content needs to be checked either by Karl Fischer or by oven-drying method and adjusted in the OB composition. Foreign matters like metallic pieces or fibers have to be screened out while making the solution. If AN has solidified, breaking it can be hazardous if metallic hammers are used. If use of melt is not possible, the next best alternative is to use uncoated AN prills (not porous prills). The use of fertilizer-grade prills is not recommended as it contains a high percentage of inert coating, which will interfere with explosives making and also cause clogging of lines. Other nitrate salts such as calcium nitrate, sodium nitrate, and magnesium nitrate if present in the AN in traces do not cause any problems.

A typical uncoated prill for use in OB making has the following specifications:

- NH_4NO_3 98–98.5%
- Bulk density 0.85–0.90 kg/l
- Moisture 0.5–0.75%
- $Mg(NO_3)_2$ 1.0–1.2%
- Inert coating nil
- pH of 30% solution 5.0
- Organic matter <1.5 ppm
- Average size (1–2 mm) 55%.

For 93% AN melt the pH should be 4.5–5.5, free of chloride, iron, and sulfate. Crystallization point is 94 °C.

An easy way of checking the AN/H_2O percentages is to measure the crystallization point of incoming material and compare against standard curve obtained by plotting of AN content versus crystallization point (Table 5.6).

If AN is to be added as an additional strength enhancer on top of the already made explosive, then porous prills would give the best results. The type of the porous prills for getting the optimum performance and energy output has already been discussed in Chapter 4.

The criteria mentioned above for checking the quality of AN also holds good for other nitrates to be used in the formulation. These are purity, moisture content, absence of foreign matter, absence of nitrite, and coatings. As for calcium nitrate, the water content needs to be determined both as free water and as

Table 5.6 Crystallization point of AN and other salts in water.

AN (%)	Water	SN (%)	SPC (%)	Crystallization point (°C)	Density (g/cc)
80	20	—	—	59.5	1.355
85	15	—	—	77.0	1.382
62	20	18	—	51.0	—
50	22	14	14	32.0	1.552

water of crystallization. Usually calcium nitrate is available commercially with four molecules of water of crystallization which has to be accounted for in the composition.

In Scandinavia a popular product easy to handle and transport containing calcium nitrate as solid is produced in the form of prills for use directly in slurry/formulation. The specification is given below:

NH_4NO_3	6%
H_2O as water of crystallization	15%
$Ca(NO_3)_2$	79%

Water: Industrial grade water can be used for slurry/water gel explosives manufacture but should not contain ions interfering with hydration of thickeners or cross-linking, or in case of emulsion explosives the action of the emulsifier. pH should be between 6 and 7, free from solids/particulate matter, nonchlorinated and Fe content should be below 0.1%, and dissolved solids should be less than 50 ppm.

Fuels: Fuels like sugar, urea, diammonium phosphate, formamide, ethylene glycol, and sulfur are all commonly used. For liquids used as fuel, chemical purity is checked by determining the boiling point and miscibility with water should be checked. For solid fuels, particle size, purity by measuring melting point, and absence of foreign matter should be checked. Particle size has an important influence on the final consistency of the product before packing and cross-linking.

Thickeners: Natural GGs are used very commonly as thickeners. Viscosity buildup in water is an important criterion for use in explosives of slurry and water gel types. In each process, formulation could be unique to a manufacturer and will require GGs tailor-made in this viscosity buildup. Final viscosity attained not only in water but in the presence of high concentration of salts needs to be checked to conform to specification.

Another important ingredient which can function as both fuel and sensitizer is AL. Its role in slurry/water gel explosives is explained separately in detail elsewhere in the book.

The role of GG in rheology of these water-based explosives, as also in determining the stability and sensitivity of these explosives on storage, is explained separately.

5.4.2
End Product Specification

Difference between slurry and water gel lies in the end product rheology, sensitivity, and shelf life. The slurry explosive has a low viscosity to enable pumping. It has a short shelf life of days rather than months and its sensitivity is low. In most cases, only boosters can initiate slurries and water resistance is also poor. The VOD is reasonably close to ideal in larger diameters. All these properties are better in water gels. The slurries are most suited to bulk operations and a compilation showing major differences is given in Table 5.7.

5.4.2.1 Development of New Formulations

The fundamentals of formulating a slurry/water gel is based, as in AN/FO and other explosives, on the concept of having a fully oxygen-balanced product which consists of oxidizer mainly AN, fuel, sensitizer, and water. The proportion of these four main ingredients is varied as per the requirement guided by process, performance, economy, and end use.

The most practical way whether you start from scratch to develop a product or modify an existing product by making changes in an existing formulation is to make initially laboratory batches of new formulation and observe its performance by subjecting it to tests; once satisfied, the formula can be upscaled. Even upscaling

Table 5.7 Comparison of properties of AL slurry and water gel.

Properties	AN-based slurry	AN-based water gel
Rheology	Flowy, low viscosity, can flow by gravity	High viscosity, semisolid, needs pressure to flow
Shelf life	Days	Months
Sensitivity	Booster	Cap-sensitive
Critical diameter in inches	3	7/8
Water resistance	Poor	Good
Postblast fumes toxicity	Tolerable	Acceptable
Usage	Mostly in bulk	Packaged
Cross-linking	Low	High
Sensitizer	Coarse AL, TNT	Flake AL powder, organic compounds, air bubbles
Density (g/cc)	1.20–1.35	1.1–1.25
VOD range (m/s)	4000–4500 in 4 in. diameter In open	3300–3800 in 1 in. diameter In open
AN content (%)	>70	45–60
Water content (%)	15–18	8–25

is done in stages from laboratory batches of say 1–2 kg to pilot plant stage of 5–15 kg and then on to a few plant batches, then one entire shift, then a few days of production. Only thereafter if the product is acceptable then it is introduced for regular production. In this way, wastage and unnecessary effort and unsafe situations are avoided. The same methodology is adopted for change in process conditions or when introducing a totally new process.

Thus it is very important to have a well-equipped explosives development laboratory and pilot plant. The pilot plant material can also be used for field trials. It is the author's experience that at least 100 tons need to be made overall and used to be able to establish the credentials of a new formula in the field. This entire process takes 4 months of time from start to finish. Thus planning for any change has to be done well in advance unless it is an emergency situation. One way of avoiding a critical situation usually caused by lack of RM totally or availability of only off specification RM is that while regular production is going on with established RM/process, other products based on other RMs are developed and kept ready for introduction whenever necessary.

It has been the endeavor of almost all development work to obtain explosives which are stronger, with higher density, and safe to produce and economical. Developments of special products may call for a different goal, for example, in case of permissible explosives, the primary criteria would be that the explosive passes the incendivity tests laid down by the statutory authorities or for underwater explosive the criteria would be yet another such as withstanding pressure desensitization.

5.4.3
Role of Aluminum in Water Gels and Slurry Explosives

It is a well-known fact that AL has been used as an energy source since many decades. AL has always been associated with energy in thermite mixtures as an intense source of heat energy. It is not therefore surprising that AL was a natural choice as an energy giver in explosives. While compositions containing self-explosive ingredients were sufficiently energetic, those without them needed to be supplemented for energetic performance through addition of AL.

5.4.3.1 Atomized and Flake Powders
The role of AL is different in different types of AN explosives. In slurries and water gels, it acts both as sensitizer and a source of energy. In case of emulsions and AN/FO, its role is limited to that of energy source only. In all these applications, the AL is used to increase the density of the explosive as well if required due to field conditions. The AL used also differ in their characteristics and are manufactured differently. The two types of AL used are atomized and flake varieties of AL powder. A summary of their important properties is given in Table 5.8.

Atomized Powder The most common form is the atomized powder. The name itself gives a clue as to the method of its manufacture. In brief, it is made by a process of atomization of molten AL in an inert atmosphere, cooling

Table 5.8 Characteristics of atomized and flake aluminum powders.

Properties	Atomized	Flake powder
Particle size	—	—
Screen analysis (%)	—	—
Retained on		
100 mesh	0.5–1.0	—
200 mesh	7.0–10.0	—
325 mesh	75.0–85.0	—
Passing through		
325 mesh	<5.0	99.0
Packed density (g/cc)	1.5	—
Apparent density (g/cc)	—	0.15
Bulking value (gal/lb)	—	0.0480
Oxide content (%)	0.1–0.2	4–6
Purity (%)	>98	92
Surface area (m^2/g)	0.23–0.98	8.3
Water covering area (cm^2/g)	Nil	10 000–20 000
Flake thickness (in.)	—	0.0000254
Granulometry	Spherical	Flat flakes
Reaction with water	Slow at room temperature	Very slow
	Fast at 60 °C	—

the droplets, and sieving and grading the solidified particles into different size fractions. The particles are close to spherical in shape and high in their purity.

In practice, the AL powder is added to the explosive toward the finishing stages and thoroughly mixed. When such an explosive is initiated and explodes, the AL takes part in an oxidation reaction releasing large quantities of thermal energy. This in turn expands the hot gases produced during decomposition of AN manyfold and enhances the borehole pressure. The rapidly expanding gases enable more work to be done on the surrounding strata than what would have been possible with an explosive of similar composition but without the AL.

It is a well-known fact that to achieve ideal detonation, the energy has to be released rapidly. The faster the release, the better is the performance of the explosive. The properties of atomized AL powder found preferable for this to happen are as follows:

- Finer particle size
- Maximum chemical purity
- Stability when exposed to atmosphere – no autoignition
- No premature reaction with water or moisture from air
- Thickness of oxide and nitride coatings on the Al surface to be low
- Absence of passivating chemicals to prevent total inertness.

It has been found that in addition to higher energy, there is an increase in the density of the explosive to the extent of 6–10% when atomized AL is added to the compositions. This increase in density is beneficial in blasting without reduction in the sensitivity of the explosive. Although at higher density there should be a loss of sensitivity due to decrease in the number of available hotspots needed for propagation of the detonation wave, the AL added compensates to some extent by functioning as a good conductor of the detonation wave by its own reactivity and release of heat energy.

The increase in density is advantageous in practical mining applications as more energy can be packed in a given space and the blast effect can be enhanced. In practice, there is a limit to the benefits that can be derived by the addition of AL and this is usually at 6–8% of the explosive composition. The reasons for this limitation could be the following:

1) Overlapping of the particles and hence available surface area for reaction is not increasing proportionately to the quantity of AL added.
2) Increase in density creating sensitivity problems requiring larger amounts of boostering in long column charges.
3) Extra cost involved at higher percentages not commensurate with the field performance.

An interesting phenomenon seen experimentally is the reduction in the VOD (5–10%) in explosives compositions where AL atomized powder was added to enhance the explosives performance. This was not seen when AL flake powder was added. The results are shown in Table 5.9.

The reduction in VOD on addition of atomized and chopped foil Al can only be explained by inferring that the reaction rate of these varieties of AL is slower than the reaction rate of the decomposition reaction of AN. Further it is thought that an

Table 5.9 Influence of aluminum powder on VOD of water gels.

Serial number	Product	Density (g/cc)	VOD (m/s)	Initiation details
1	Nonaluminized water gel	1.10	4000	3 in. diameter unconfined with booster
2	1. With 3% atomized aluminum powder	1.12	3750	Same as in 1
3	1. With 3% aluminum flake powder	1.10	4300	3 in. diameter unconfined with no. 6 detonator
4	1. With 3% atomized and 3% flake powders	1.12	4400	Same as in 3
5	1. With 15% MMAN	1.12	4200	Same as in 1
6	1. With 30% MMAN	1.12	4500	Same as in 3
7	6. With 3% atomized aluminum powder	1.12	4200	Same as in 3
8	1. With 3% aluminum chopped foil	1.12	3500	Same as in 1

induction period exists before the AL starts reacting and acts as a heat sink initially before participating in the oxidation reaction. This induction period is less and less for powders with increasing surface-to-volume ratio and lower inert covering on the AL powder. By the same token, the explosives with flake powder do not show any reduction in VOD.

Considerable effort has been made to see whether the AL in its atomized form could sensitize a non-cap-sensitive explosive to such an extent that it can be initiated by a No. 6 detonator. However, so far attempts in this direction have failed. It would indeed be a breakthrough if cap-sensitivity can be achieved only by addition of atomized powder and will mean a substantial reduction in the cost of manufacturing cap-sensitive small-diameter explosives. One possible explanation could be that the atomized powder even of the finest variety does not have surface area large enough to provide substantial contact between the metal and the air bubbles already entrained in the explosive. The particulate thickness also precludes instant heat transfer between the hotspots and the metal necessary for initiation and propagation.

AL atomized powders are found to affect the stability of water containing AN explosives during storage due to the reaction between AL and water. Details of this reaction are discussed later in this book 5.4.3.2.

Apart from AL in its atomized form, other types of Al such as chopped foil and granules have been used as energetic fuel, but the most consistent results have been obtained by spherical atomized powders [7].

AL Flake Powders as Sensitizer   AL flake powders are well known in the paint industry [8] and are characterized by their large surface area. They are obtained by ball milling atomized powders or foil for 10–12 h in liquid medium. Special additives are added while ball milling to obtain the desired surface. Finally brush polishing of the dried flake powder in rotating drums is done to obtain smooth surface on the flake powder. The flake powder can also be obtained by ball milling or stamping process in the dry state. But most modern processes use ball milling in liquid media and then driving off the solvent in a Stokes drier to the desired level of dryness. The flake powder is sieved to the desired size. The operations are conducted in inert atmosphere and the final exposure to air is done under controlled conditions and isolation as there is a tendency for the freshly made flake powder to catch fire. The surface area and its properties desired are obtained by varying the parameters of the ball milling process such as speed, size, and number of balls in the ball mill, period of the ball milling process, and amount of stearic acid used.

It was not until 1968 that Van Dolah [5] described in his work at the USBM the ability of AL flake powder of a certain specific surface area to impart such a degree of sensitivity to a water gel that it could be initiated by a No. 6 detonator. He further described several formulations where similar results were obtained and confirmed the earlier findings. Based on this, further work was carried out by Gehrig and Mahadevan (1970, Atlas Powders Ltd, USA) [2] and commercial production was established for these types of AL-sensitized water gels in small diameters which were cap-sensitive and could compete with NG-based gelatins.

The addition of AL flake powders ranging from 8000 to 22 000 cm^2/g was found to impart cap-sensitivity. The percentage added was 3–6%. An optimum dosage of 3% of AL flake powder with 12 000 cm^2/g surface area was established. The surface area was measured by means of a procedure of water coverage used in the AL paint industry and checked for reliability using Fischer sub-sieve-size analysis and BET method. The AL flake powders coated with a hydrophobic coating of stearic acid were able to retain close contact with the air bubbles already in the explosive matrix as also bring in certain amount of air into the explosive because of their low bulk density. (For instance the density of the water gel matrix at 1.20 g/cc fell to 1.12 g/cc when 3% AL flake powder was added and mixed.) The flake powder then acted as conducting and propagating facilitator for the detonation phenomenon by transferring the heat energy generated by the adiabatic compression of the air bubbles along the path of the detonation minimizing any side losses. The flake powders, because of their thin surface dimensions, are also highly reactive when subjected to thermal shock and helped in the continuous propagation of the detonating wave. This property has been made use of in designing cap-sensitive water gels capable of being used in small diameters (1 in. diameters).

The AL flake powders are manufactured worldwide by many of the large AL companies. The exact requirements of the explosive manufacturer can be arrived at only by close interaction with the supplier and by carrying out trials with samples initially before fixing the final specification of powder for bulk production. This procedure is highly recommended by the author who in his experience found that no two AL flake powders obtained from two different sources but of identical specification were same in surface properties and could behave differently as sensitizers in explosive. The quality of stearic acid and solvent used during ball milling is found to affect the surface properties and hence purity of these ingredients must be of a high order and also frequently they have to be changed because of in-process contamination. The solvent needs to be removed fully as any residue affects the surface tension of the flake–air bubble interface in such a way the air bubbles are not stable and either collapse or become larger due to coalescence affecting the sensitivity of the explosively badly. The effect can be such that cap-sensitivity is lost completely. The rate and method of removal of solvent also needs to be controlled so as not to damage the surface coating of the flake and in this regard rotary vacuum dryers have been found very suitable. The dry powder will need careful handling as it is pyrophoric and catches fire on exposure easily.

5.4.3.2 Aluminum Water Reaction

Apart from oxidation to Al_2O_3 during detonation adding to explosive energy, there is the disturbing phenomenon of AL reacting with water in the water gel in storage and disrupting the structure and enabling the air bubbles to escape. This reaction cannot be fully arrested but slowed down to be ineffective during lifetime of the explosives. The reaction proceeds faster at higher temperatures such as 50 °C and in alkaline medium (pH > 8) and in the presence of NaCl, sodium perchlorate, and

Al_2O_3. The reaction proceeds very slowly or not at all with flake powders properly coated with stearic acid and hence gassing phenomenon (release of H_2 with heat) is more common with atomized powder. It is now possible to buy atomized powder also coated with 0.5–1.0% stearic acid for protection against reaction with water.

The reaction of Al with water as in

$$2Al + 6H_2O = 2Al(OH)_3 + 3H_2 \tag{5.1}$$

is thermodynamically favorable enough to proceed at room temperature, but this does not happen due to the fact that AL is covered by a layer of Al_2O_3 in atomized powder and in case of flake powder with stearic acid or stearate as well. If there is a coherent and adhesive covering, then it is very difficult for water to come in direct contact with AL metal and reaction (5.1) does not proceed. This can be checked by detecting if any H_2 is evolved. Temperature dependency of the reaction can also be determined by observing rate of H_2 evolution at higher than room temperatures. In practice in slurries and water gels, the presence of water and salt can make the reaction proceed faster.

In fact during studies carried out for AL/H_2O reaction as a possible source of hydrogen, it was found that the reaction could be accelerated by three types of promoters: (i) hydroxide, (ii) oxide, and (iii) salt.

The reactions are

$$2Al + 2NaOH + 2H_2O = 2NaAlO_2 + 3H_2$$
$$Al_2O_3 + 2H_2O + 2NaOH = 2NaAlO_2 + 3H_2O$$
$$Al(OH)_3 + NaOH = NaAlO_2 + 2H_2O$$

In some production process of AL powders, they are exposed to N_2 and could have a layer of Al nitride. Even this reacts with water/NaOH

$$2AlN + 3H_2O = Al(OH)_3 + NH_3$$
$$ALN + NaOH + H_2O = NaAlO_2 + NH_3$$

All the above reactions are accelerated as the temperature rises. The temperatures when AL comes in contact with the rest of the explosives during production can range from 40 to 70 °C and the composition contains H_2O from 10% to 25%, although they are in bound form with GG.

The stability of the explosive where AL powder is added as an ingredient depends on the quality of the Al_2O_3/stearic acid coating on the AL powder. They have to be uniform and adequate. Then only the explosive will not be subjected to the negative effects of Al/H_2O reaction, which are as follows:

1) AL is prematurely consumed, does not contribute to energy, and sensitivity of the explosive.
2) It disrupts the gel structure and causes water separation, and allows entrapped air bubbles to escape.
3) It produces intense heat and H_2 which are both dangerous in confinement.

The explosive becomes powdery – all the H_2O having been consumed. In this state with some unreacted AL still left in the composition, which can be more sensitive than in the original state to impact and friction and has to be handled carefully as it can still function as an explosive. I have experienced an instance in a magazine in the Middle East where in the water gel the AL/H_2O reaction took place because of storage at very high temperatures of 60 °C and the explosive became powdery. The latter was found cap-sensitive, while the product originally was only booster-sensitive.

As seen from the probable reaction mechanism, it is essential to prevent hydroxide promotion of the Al/H_2O reaction. This is done by keeping the pH at 5.0–5.2 at all times during mixing and in the finished state. The oxidizer blend when made could have pH beyond 6. This should be brought down by addition of acid (acetic or nitric) to the desired range. An important point to note is to buffer the oxidizer solution by using diammonium phosphate so that even during storage the buffering keeps the pH in the desired range. This step, coupled with use of AL powder of good coating, should ensure adequate shelf life to the water gel even if due to some other reason the gel formed becomes weak and some water is exuded.

The surface property of the flake powder apart from its covering area is important for maintaining the sensitivity of the explosive. The surface has to be flat and thickness of the flake should be uniform. In practice the surface has some wrinkles and indentations obtained during the ball-milling process. It is much more nonuniform in dry stamped powders. The brush polishing of the flakes in the presence of stearic acid not only removes the wrinkles to a great extent but also ensures uniform hydrophobic properties. This is very necessary for the air bubbles in the matrix to remain in contact with the Al flake thereby facilitating the conduction of heat developed during the adiabatic compression of the air bubbles. The flake being very fluffy has a low bulk density and the flake structure divides the air/gas bubbles present in the matrix to finer size bubbles (micron size). For example, a standard water gel at cup density 1.20 dropped to 1.12 g/cc on addition of 3% AL flake powder of bulk density 0.12 g. It is also important to have a uniform flake thickness so that thermal conductivity from air bubble to AL is uniform.

In general the thinner the flake, the more reactive it is and better its use as a sensitizer provided it is well-protected against action from water during the lifetime of the explosive. One manufacturer has measured a film of AL of density 2.58 g/cc, weight 1 g, and having a thickness 0.000254 cm to cover an area of 15 740 cm^2. This powder was found to be a good sensitizer with 2.5–3.0% stearic acid as coating.

Any trace of mineral oil or solvent is found detrimental to the sensitivity of the water gel explosive since it influences the size and stability of the air bubble structure by changing the interfacial tension. The volatiles on the AL flake powder have to be removed completely by vacuum drying.

5.4.3.3   Important Tests for AL Powder for Use in AN-Based Water Gel Explosives

Particle size: Done by standard method of sieve analysis in the dry state for atomized powder and by wet sieving (in solvent media) for flake powders. Finer size particles are more reactive in general.

Volatiles: Applicable to flake powders must be less than 0.1%. Determined by slow drying method with careful watch for ignition and loss of weight to be measured in sensitive balance. Determination of volatiles can also be speeded up by drying under vacuum.

Fat (grease) content: Refers to stearic acid coating on flake powder. Method employed uses acetone to extract the stearic acid from the residue obtained after dissolving completely the AL powder in hydrochloric acid. The acetone extract is then evaporated and the stearic acid left behind is weighed and percentage on AL calculated.

Oxide content: It is important in estimating the purity and energetic value. After removal of grease using solvent, metal is taken into solution, precipitated as hydroxide, and ignited and weighed as oxide. From this, metallic AL content can be calculated. Making AL react with HCl gas, and converting all AL to AL chloride and estimating the oxide left behind (does not react) is another way of obtaining AL content of the sample.

Density: It is measured usually as apparent density. The weight in grams occupying a known volume (100 cc) in a fully filled measuring cylinder without tapping after standing for 5 min is divided by 100 to arrive at the value of apparent density in gram per cubic centimeter. Apparent density of flake powders is very low (0.12–0.15 g/cc). The finer the flake powder, lower is the apparent density.

Leafing value: This number is used mainly in the paint industry. It is the ability of the AL powder when swirled in certain liquids to come to the top and form a continuous surface. This test is useful to the explosives manufacture in that it is desirable to have good leafing ability in the flake powder being used. However leafing alone will not determine the efficacy of the AL powder for sensitizing the explosives formulation.

Granulometry: It is important to find out the shape of the AL being used. Atomized powder should be spherical or slightly tear drop shaped. Flake powder should be as flat as possible. Granulometry is obtained by examining the AL powder under a microscope (high-power laboratory-type) for routine checks. For more detailed information, the powder is examined by means of scanning electron microscopy.

Water covering area: It is one of the most important properties of the AL powder giving a clue as to its effectiveness of the AL flake in the explosive.

In general as proved by experiments, the higher the water-covering area the better the powder works as a sensitizer in the water gels.

Water-covering value serves also as a means to calculate the total surface area of the flake as well as its thickness. The water-covering value can be used as a routine test for accepting or rejecting a consignment of AL flake powder. The test takes about 15 min and gives reproducible value. The method is based on the ability of the flake powder to spread out in a thin film (one flake thick) on the surface of water.

The test is carried out with a simple apparatus. A rectangular tank holds the water and two glass pieces slide along the rim. Water is filled to the top. A known weight of flake powder is added to the water in a wide area, and with a glass rod the clumps are broken up and the flakes allowed to come to the surface and form a film. This process requires some skill and patience on the part of the tester. Thereafter the two glass plates are moved in tandem from opposite direction in and out so that first a thick film is formed, which is then extended as far as possible without a break or wrinkle. This is the maximum area that can be covered by the flakes without overlap of particles. This area is the water-covering area for the weight of the flake taken and hence the same is expressed as square centimeters per gram. The value ranges from 8000 to 25 000 for most of the flake powder manufactured in different fineness and apparent density. Reproduction of the values is good once the tester develops the required skill. The dimension of the tank and the weight of powder tested can all be standardized by trial and error. Usually the tank is about 50 × 15 × 2.5 cm. The inside is smooth. Glass slides are longer than the width of the tank. Edges of pan are coated with paraffin to prevent water from flowing out and this also helps in smooth sliding of the glass plates.

From the water-covering value, weight of the materials taken, and density of AL taken as 2.5 g/cc, thickness of flake can be calculated.

Surface area: It is also of interest to the explosive industry as it relates to activity of the AL in the detonation. As a first approximation, twice the water-covering area gives us a value for surface area. But this is only an approximation (albeit a good one) since it does not take into account the possible voids between flakes in the film on the water. They may be about 5% of the covering area. Other factors to be considered are edges, irregular contours, and wrinkles. All these lower the real surface area to some extent. BET method of adsorption is used when very accurate measurements are to be made. There is also the possibility of using air permeation method but, this does not work well for flake powders due to layering but works well for atomized powder.

If the BET surface area is divided by twice the covering area corrected for voids, a roughness factor can be obtained which gives a measure of the irregularity and wrinkles, pitting on the surface obtained during ball milling and not removed by polishing. Flake powders have 1.3 as roughness factor as against atomized powder which has 4.5 depending on the fineness.

Stability Test   The reaction of AL powder with water, hydroxide, and salt disrupts the gel structure and lowers the performance of water gels. In extreme cases, the heat and hydrogen evolved can lead to hazardous situation. Hence it is important to use AL powder, which remains rather inert as far as the Al/H_2O reaction is concerned, though remaining active for release of energy and taking part in the combustion process without any delay.

To measure the stability of the AL powder, a simple laboratory set up for measuring the volume of gas (H_2) evolved is used. A known weight of the AL powder is mixed with H_2O or with the oxidizer blend being used for manufacture of the explosive and kept at constant temperature. The gas evolved after specific periods of time is measured and compared with standard as also theoretical values at normal temperature and pressure. All comparisons have to be made at the same pH for the reaction media.

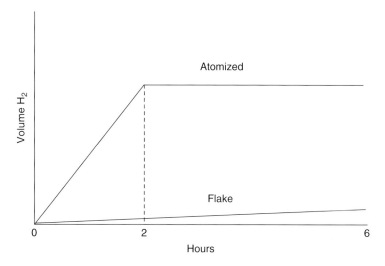

Figure 5.2 Stability of aluminum powders in water/NO · OH at 60 °C, pH = 7.5.

Usually the reaction at 25 °C in H_2O is hardly measurable, but we need to know the reactivity under practical storage conditions. Hence the H_2 evolution is measured at evaluated temperature such as 60 °C and measured every 1 h till no more or very little gas is produced. The first 2 h is usually critical. Atomized powder with no protective coating of stearic acid reacts very fast but a well-coated AL flake powder does not react at all. As a further step in making the test more severe, the effect of alkali/NaCl can also be measured by keeping the pH above 7 and also by taking NaCl in the solution in which the Al powder is mixed. The explosive manufacturer can modify the parameters of the test depending on the shelf life required for his product and the type of AL powder one wishes to use.

The reactions involved are given in detail in Section. ... Experience shows that this test is very useful in weeding out any unstable powder if the results are properly evaluated.

Figure 5.2 shows evolution of hydrogen gas with time for atomized and flake AL powders.

Reactivity of the AL Powder Oxidation of AL powder in air to form Al_2O_3 film is inevitable in the first instance, but further oxidization is prevented by the formed oxide layer. The protection is further enhanced if the AL powder is also covered with stearic acid. However under certain conditions, this AL is still required to ignite and release heat rapidly during detonation. Thus both the oxide layer and the stearic layer should be minimal enough to protect but not so much as to make the powder inactive for explosive purposes. Experience has shown that oxide content limited to 0.1–0.2% by weight gives the best results but is not easily attainable during manufacture. It is more practical to err on the safer side of not having a too active powder, which catches fire on its own (autocombustion) immediately after manufacture. Hence 0.5–1.0 % by weight is usually the oxide content aimed for but seldom achieved.

Storage and handling of AL powder is one of the most important concerns for the manufacturer of explosives. The desire that an additional hazard should not be created when AL powder is used, as an RM in production of explosion is understandable but not totally realizable. However the hazards associated with use of AL powder can be significantly reduced by following the recommendations and guidelines [9] issued by the AL association (TR-2). The recommendations are based on the general principles of protection and avoidance of creating friction, static electricity discharge, formation of dust cloud alone, or in combination with other ingredients, fire, and contact with water. (The recommendations are reproduced in Chapter 8.)

5.4.4
In-Process and Finished Product Checks

5.4.4.1   Oxidizer Blend Composition
This is checked quickly and effectively by measuring crystallization point and comparing with the standard for OB of identical composition made in the laboratory with carefully weighed contents. This simple method consists in cooling gradually a sizable sample (100 cc) of OB from its holding temperature (80–90 °C) and noting the temperature at which crystals emerge from the solution (seen as a hazy cloud).

In the plant even when the required composition is followed, due to bulk quantities being added in preparing OB of 20–30 tons at a time, the composition can deviate due to weighing errors inherent in the weighing equipment used as also evaporation and loss of water due to prolonged storage under hot conditions. Thus before actually using the OB, it should be checked for conformity to composition and variances from standard corrected by adding make up H_2O or addition of oxidizer salt. Density of the OB is another easy to measure check and can be again compared with standard densities of that composition. Care needs to be taken when using $Ca(NO_3)_2$ when making OB because of the water of crystallization and its contribution to the water content of the OB.

If crystallization point and density vary significantly, one needs to check the actual blend composition by using methods in the laboratory for driving off water and measuring the weight of residue, and estimate AN directly by titration after reacting with formaldehyde. It is best to check the crystallization point for AN/H_2O first and then after each addition again. Effect of other salts on the crystallization point can be seen. pH of OB is another important parameter to be checked and controlled. It is usually kept between 5.2 and 5.6. Buffers are used for keeping the pH in this range throughout production and in storage.

Temperature should be checked online/offline and maintained at desired level (80–90 °C). Volumetric or weight check of OB drawn into blender if batch process is used needs to be done for every batch. In case of continuous process, flow check should be done continuously using flow meters. Required degree of automation can be introduced as desired. OB should be observed physically for any contamination from anticaking inert coatings on AN and removed through filters. More visible contaminants like plastic/jute fibers can be removed physically.

5.4.4.2 Solid Ingredients
They need to be checked against their specification and their weight should be checked before addition to blender for batch process. In case of continuous process, addition through screw feeder is done. The screw feeder needs to be calibrated for flow against speed (rpm) for each of the ingredients as this will change due to difference in bulk density. Solid ingredients would also require screening to remove foreign objects.

5.4.4.3 Liquid Ingredients
Liquid ingredients such as fuel (ethylene glycol), cross-linker solution, and gassing solution need to be checked either by weight or by volume or flow so that only the correct quantity as per formula goes into the mixer. All these RMs need to be checked prior to addition for acceptability as an RM. Cross-linker chemical composition has to be checked in the laboratory. Purity and effectiveness are checked before acceptance.

5.4.4.4 Mixing
Mixing time, hydration time, aeration time, speed of blender, and rotational direction need to be followed as per manual, and data entered into the log sheet. Unusual observations if any also need to be recorded. Gel formation and viscosity during process need to be observed by the operator to detect any usual phenomenon like low hydration, bad dispersion of gum, and so on.

The mixing process aims to build a uniform stable product. To this end, the viscosity of the mix before solids additions in batch process needs to be checked. In case of continuous operation, it is more difficult and online sampling and checking is the only way. The viscosity of most commonly used thickeners follow non-Newtonian laws. Their viscosities can be accurately measured using a Brookfield rotational viscometer. For plant checking purposes, a flow check apparatus is used. The time taken for the viscous mix to flow through an orifice at the bottom of a funnel holding about 1 kg of the material is an indication of the viscosity to be compared against standard arrived at in the laboratory/pilot plant for a standard batch of identical composition.

Cup density of the mix at several stages is needed to keep track of the process in batch mixing. In case of continuous process, final density after gassing is the only check. In batch process, density of the mix after hydration and after mechanical whipping of air into the gel, after addition of AL, and final density thought the packer are all checked. Cup density is determined by filling to the brim a cylinder of known volume, determining the weight of the mix inside and calculating W/V to give cup density in grams per cubic centimeter. Tables giving density for a specified cup weight with the material can be developed for quick determination of the cup density of the mix.

5.4.4.5 Packing

During packing, weight inside the cartridge is checked very often and variations outside the process limits adjusted. It is obvious that change in density of the product will affect its weight in a space of constant volume and hence average weight needs to be tracked.

Cartridges have to be carefully inspected for sealing defects and the sealing device adjusted.

Cardboard cases need to be tested before acceptance for bursting strength, grammage, waterproofness, and stitching. In the plant before packing and sealing, printing on the case should be checked against the specification before dispatch to magazines. Glue used are usually either silicate-based or starch-based and need to be checked against specification and for efficiency before release to the plant.

Stacking strength of cases is determined by looking at stacks of cases with packed material (6 high) in storage over time. There is reasonable correlation between the properties of fiber board such as the bursting strength of the board used in making the cases and its stacking strength. Explosives manufacturer has a limited option of changing style and strength of the packing as it is governed by transportation rules and regulation, but still enough scope is there for continuous improvements in strength at optimum cost.

5.4.5
Performance Tests

VOD determination as described earlier in Chapter 3 of this book is a good measure of the product quality. This is usually done for water gels after 5 days of manufacture as release test. This period enables the cross-linking to be almost complete and the air/gas bubbles are fixed in place. The gel is also set so that it can withstand transport. Problem products and some regular ones can be stored and their VOD in storage followed. Any deterioration in VOD is an indication of the deterioration in the explosive properties. Similar information though not so reliable can be gathered by plate dent test. Weight strength tests can be checked by lead block expansion for cap-sensitive water gel. However as a routine test, determination of VOD and cartridge density of product is the basis for release to customer. Cartridge density can be obtained by measuring loss of weight in water and using it for calculating the cartridge density. Otherwise material can be carefully removed and cup density determined and compared with specification. Any abnormal increase in density over a short period is an indication to be watchful for low shelf life even though release VOD test may indicate that product is within limits.

5.4.6
Safety Tests

These are done to determine the sensitivity of product to static, friction, and impact. Whenever new ingredients are introduced, safety tests need to be performed to know their effect. These are not release tests. In case of unusual incidents, residual material is subjected to these tests for obtaining insight into probable cause of the incident.

5.4.6.1 **Gap Test/COD**
These are also release tests especially for small-diameter, cap-sensitive products performed in the open. Two cartridges are placed at a known fixed distance end to end and while the donor cartridge is initiated, the acceptor is watched for full detonation. If yes, then the product has passed the gap set. Using the Bruceton up and down method, gap distance can be obtained for 50% pass and used as standard. But in release tests, fixed distances are used for pass/fail depending on the diameter of the cartridge being tested. In confinement and in a borehole, the gap-sensitivity, that is, ability to jump a gap, is much higher than in the open. Gap test is a very useful research tool to find the effect of ingredients with different properties such as metallic powders and self-explosive ingredients. Gap tests of stored products can also detect the effect of aging of the explosive. For most water gels aging starts from 3 months onward, but the explosive may still be acceptable in performance even after 1 year depending on the composition and process used for manufacture and the storage conditions.

5.4.6.2 **COD**
Continuity of detonation (COD) is another test that can also be used for release as well as to judge the condition of explosive during its shelf life, especially for SD cap-sensitive produces. Here four or five cartridges are kept in contact with each other in a rolled paper tube to keep them in position. From one end the explosive is initiated, and it is observed whether the shock (detonation) has propagated throughout all the cartridges (a length of approximately 1 m). VOD measured in the last cartridge can give additional information.

5.4.7
Storage Tests

Physical observations, apart from performance tests, are important part of judging the condition of the explosive gel.

Gel condition on observation should not show any signs of separation seen as water/salt solution coming out of the explosive gel. This, if noticed, is an indication that the rigidity of the gel has been compromised. This will surely affect the sensitivity of the explosive to initiation as not only the composition's oxygen balance might have got affected but even the incorporated air bubbles might have escaped or coalesced to become large pockets which are less prone to adiabatic compression and heat transfer.

In certain instances, the gel might have become powdery due to premature AL/H_2O reaction and the heat generated has evaporated the rest of water. Such a powdery gel needs to be handled carefully even while testing. Salvage or destruction is decided on the basis of these tests (VOD, sensitivity to initiation, and gel condition).

All products which have prematurely gone bad are prime candidates for examining and understanding of the causes for such ill effects, and each and every ingredient as well as process will have to investigated and eliminated as probable cause to arrive at a final conclusion. Thereafter taking steps to prevent such happenings again will assume priority. Weak gel formation, weak cross-linking, and accelerated AL water reaction are most common causes of gel deterioration in storage. Storage temperature plays a very important role in all the above effects contributing to gel degradation which is accelerated by increasing temperature. Storage at $60\,^{\circ}$C is substantially disruptive as compared to storage at $20\,^{\circ}$C. Cycling of temperature is another serious cause of deterioration of gel since AN partly crystallizes out in the slurry/water gel as the supersaturated solution formed at higher temperature gets cooled. These AN crystals are subject to phase transition changes at $32\,^{\circ}$C and its ill effects (presented in detail under AN in ANFO in Chapter 3 of this book) unless additives preventing such phase changes are present.

Based on these facts, storage stability tests have been designed to evaluate the usefulness of a formula for making a stable slurry/water gel. These tests are long-term stability evaluations and hence are more useful for water gels rather than for slurries.

In the most commonly used test, the explosives cartridges are kept in an oven whose temperature is controlled by hot water circulation. Electrical heating is avoided for safety reasons. The oven itself is kept in a protected concrete structure housed in a room constructed according to norms used for an explosive magazine building. The oven with the explosive is kept at $60\,^{\circ}$C for several days. Periodically the product is removed and observed for any change in condition. It is essential to keep the oven on at the desired temperature continuously so that the explosive also attains the temperature of testing and remains at that temperature throughout the period of the test. Correlation between shelf life under normal storage conditions and in hot storage is possible by noting the time at which the product starts decaying and if large number of tests are done and data obtained, it will be possible to predict the relationship between the normal shelf life and life in hot storage. Authors experience with water gels with and without AL showed lesser stability for the former. In general if an explosive can withstand 4 weeks of hot storage without degenerating, it is most likely to withstand at least 4 months of normal storage in tropical weather conditions. In colder climates at least 6 months shelf life can be expected [10].

The effect of temperature cycling done according to the test described in the earlier chapter has not yielded a clear relationship between cycle life and normal storage life in case of water gels.

Cold temperature affects the gel condition (not stability) and performance of most water gel compositions much more in the short term than storage at elevated temperature. At less than 15 °C in most common water gel/slurry formulations, the major part of the AN crystallizes out of the solution layer and this is enough to upset the compositional balance and homogeneity resulting in low sensitivity to initiation.

The crystallizing AN could also destroy the microstructure of the air bubbles thus affecting sensitivity and propagation. The effect of cold temperature is less pronounced if the microbubble structure is provided through synthetically formed microballoons, though this does not compensate for AN coming out of solution and affecting the performance. Whenever explosives are likely to be used in winter at cold temperatures below 15 °C, it is best to have a formula which works at these temperatures efficiently. The explosives must be tested by keeping it at low temperature in a freezer for at least 2 h or leaving it overnight in a magazine in winter so that it attains a low temperature. Then it is taken to the testing site in an insulated box to prevent it warming up and tested for booster/cap-sensitivity, VOD, and COD. The temperature of the explosive inside the cartridge is measured and noted just before it is initiated.

The initiation sensitivity, VOD, and propagation ability of most slurries/water gel fall with temperature. Unless special formulas are used, they fail to function between 10 and 15 °C in unconfined testing. Luckily most boreholes, especially the underground ones, are much warmer than surface air temperature and hence many explosives, if allowed to attain borehole temperature before blasting, can function even if they have failed on surface. However it is best to keep an adequate margin of functionality to reduce risk of failure and avoid all the problems associated with it. Special formulas can take the low-temperature functionality down to 0 °C, but these formulas could be more expensive.

5.4.8
Gel Condition Evaluation

Physical examination of the explosive is done to get an idea of the state of the water gel after it has been given sufficient time to cross-link, usually 5 days. Finger pressure or probing with a glass rod is done to obtain a measure of the gel strength. Numbers 4 to 1 are given in the descending order of the gel condition – No. 4 for very good and No. 1 for very bad. A good gel is one which springs back after getting depressed. A very bad gel is one which allows the finger or plunger right into the body without any resistance.

Gel rheology is also important to judge whether hydration and cross-linking have taken place. Both these can be gauged by the finger depression test. In addition, the gel is pulled manually if it comes away in pieces; the hydration of the gum was not completed in the process. An ideal gel is one which is elastic, firm, and does not break into pieces when pulled.

Gel rigidity test as described by Goring and Young [11] is a good scientific way of evaluating the gel strength. The test is less person-dependent since it consists of mechanically driving a plunger by a motor at constant speed through the gel kept in the pan of a balance and the movement of the plunger with respect to time is measured and used for obtaining the gel strength.

Some of these tests described are subjective tests depending on the judgment of a person, but with experience they can be used in most situations to evaluate the explosive gel condition and estimate the remaining useful shelf life. Of course confirmation is always best done by actual firing tests.

5.4.9
Waterproofness Test

Water gels are designed to withstand ingress of water to a certain extent and it is important to know the extent of this resistance. As in AN/FO, the effect of flowing water is more severe than static water in a borehole. The tests described under AN/FO in Section 4.4.4.7 of this book can be used for water gel also.

The tests will need to be conducted for longer period, sometimes for days, to obtain the end point of the test, that is, when total disintegration of the water gel has taken place. In practice most slurries which are less water-resistant than water gels but more than AN/FO are loaded and blasted within 2 days maximum in wet conditions and if longer sleeping time is required water gels in packages are used. Thus waterproofness tests are accordingly adjusted in duration. For the slurries, 48 h maximum and water gels up to 5 days are the periods practiced. If the results are satisfactory at the end of these testing periods, the formulas used are quite adequate for most blasting situations where water has been found in the boreholes.

5.4.10
Effect of (Hydrostatic) Pressure

Since slurries/water gels are used, in deep LD boreholes, at the bottom, the explosive will be subjected, apart from the leaching effect of water, to pressure from the column of explosives extending up to the top as also hydrostatic pressure from the water in the borehole which has been displaced upward. This increases the density of the explosive at the bottom since the explosive material is compressible being soft. Further if chemical gassing or mechanical aeration is used, the air bubble structure could get disturbed in a negative manner. If the gel is weak, the smaller bubbles could coalesce to form bigger ones leading to reduction in the sites for hotspots which in term will reduce the performance. In extreme cases, initiation sensitivity gets lowered to such an extent that additional boostering is needed to initiate the charge. It is noticed very often in underwater blasting.

A simple apparatus can be rigged up in the testing site to simulate overpressure on the explosive cartridge and conduct firing tests thereafter. Test apparatus used is given in Appendix A of this book.

In some cases, repeated application of hydrostatic pressure is necessary to obtain failure as a certain amount of spring back to normal condition is possible in a good water gel once the applied pressure is removed.

5.5
Process Hazards (Dust Explosions/Fire Hazards/Health Hazards)

During the manufacture of slurries/water gels, the hazards encountered during process have to do less with its explosive properties than with other types of hazards. The process of water gel manufacture starts with a nonexplosive mass and sensitizes it, whereas in the manufacture of dynamites one starts with an extremely sensitive mass initially and brings the sensitivity down by adding more inert materials. Hence till the last step of aeration/and addition of sensitizer like aluminium flake powder, the hazards are quite less even if the material is treated harshly. On the other hand, one has to guard against dust explosion both when adding GG and AL powder. One tends to ignore the former and concentrates only on the latter but fine GG particles suspended in air are also known to be sources of explosion.

As far as dust explosion and other hazards encountered when AL powders are used, much investigation has been done and details presented in literature [8, 9].

The dust explosions are generally held to occur more frequently and at lower concentrations for finer particles with larger surface areas. Thus AL flake powders show a greater tendency toward explosions in air than atomized powders. In case of GG, drier and finer the gum powder, greater is the risk. As low as 40 mg of AL powder per liter of air is enough to have a potentially explosive mixture. The manufacturing process for slurries and water gels for convenience sake involves addition of AL powder/gums/sulfur mixture and this combination is very prone to dust explosion as each of the individual component itself is a contributor to the hazard. The most practical way is to avoid using these powders in the dry form, especially as a mixture, and use compatible liquids to wet them before adding to the oxidizer blend/matrix. If adding as powder is inevitable, the area (floor) where the mixer is situated should have free-flow natural air ventilation .This is not difficult as the mixer floor usually is well above the ground level (second or third floor). Air movement carries away the suspended solids into the open, dilutes the concentration of active material below hazard level, and ensures a safer operation. Earthing of equipment to prevent static hazards is a must and also goes a long way in reducing explosion risk.

Fire hazard is always present, again because of the presence of fine metallic ingredients used like AL, iron powder, nonmetallic fuels like sulfur, liquid fuels like diesel oil, natural products like flour, and various types of gums. Appropriate fire protection and firefighting measures need to be taken. They may be different for different materials. Natural product fire can be quelled by water or chemical foams. AL fires are to be treated according to guidelines issued after much study by The AL Association [9]. These prohibit use of water, liquid, and foam types of extinguishers. Shutting down of all equipment, cutting of electrical power, avoiding air current drafts and physical disturbance of the burning area, and allowing the fire to resolve in the initial stage itself by use of dry extinguishing agents are recommended steps. Once oxide layer is formed, the fires tend toward self-extinguishment.

5.5.1
Slippery Floor

It may appear as a minor problem but when natural product thickener, viscous, and oily fuels are used, the floor can become a hazard that should not be ignored. The GG powders falling on the floor tend to absorb moisture due to hygroscopicity and form a very slippery film (very difficult to dislodge) which can prove to be dangerous to moving personnel. Such films are best removed by scrubbing and washing with plenty of water.

5.5.2
Health Hazard

Ingestion of dry materials present in air can cause short-term breathing problems, which resolve by themselves once the human moves out of the contaminated area into cleaner surroundings. Long-term effects of being exposed on a daily basis to powders/dust clouds of AL powder [12] and GG have been subject of studies and it has been opined that no health hazards, beyond what a human incurs by atmospheric pollution, exists. In fact some studies have shown that these two materials when ingested tend to remove silica from the lungs and both of them do not accumulate in the body.

Still as a matter of precaution, manufacturers use some steps to avoid even remote possibility of contamination:

1) Use of dust protection mask in sensitive areas
2) Rotation policy for employees working in these areas
3) Use of milk/lactose tablets as special nourishment for those involved
4) Use of completely closed systems for handling of these powders from incoming to addition
5) High-pressure water washing and air cleaning of hair and clothing before workers end the shift and go home.

5.6
Role of GG

GGs have been known to mankind from ancient times as beneficial in diet. This has been later confirmed by modern medicine. It was also found that it had a very good property of thickening rapidly in water even when dissolved solids were present and could form a stable gel by cross-linking with some elements. Hence it was and is being used extensively in the food industry, in pharmaceutical products, and in the printing industry [13, 14]. It was the discovery of slurry explosives and the need to thicken the product that opened up the area of GG useage in the explosive industry. With the rapid growth of all types of slurries and water gels, the requirement of GG increased also and specific needs of the explosives manufacturing process had to be met by modifying its viscosity, cross-linking, stability under different conditions. With the introduction and growth of emulsion to a certain extent the demand for GG in the explosive industry has lessened.

5.6.1
Application in Water Gels and Slurries

As already described under process, supersaturated solution of AN and other oxidizer salts is thickened by adding GG and fully dispersing it so that viscosity is built up rapidly and cooling the mass thereafter. During cooling the salts crystallize out, but in the presence of the viscous solution the crystals are fine (small) in size and do not agglomerate as a film of GG in H_2O acts as a physical barrier between the crystals. This is particularly useful since AN tends to bridge and cake up when the particles in the form of crystals come in close contact with each other.

GG has also the property of entrapping in its body air bubbles generated through mechanical agitation and gas bubbles released by chemical reaction. The stability of the bubbles in the thickened matrix is enhanced by cross-linking using certain chemical compounds.

By the action of the cross-linker, the viscous matrix becomes a nonflowy, semisolid material where the air bubbles are (fixed) immobile to a considerable extent. This phenomenon is of great use to make the explosive mixture acquire greater sensitivity due to the presence of hotspots in the form of the air bubbles. In fact the enhancement can be of such a magnitude that a booster-sensitive mix can be made cap-sensitive by increasing aeration.

GG in order to function usefully in explosive manufacture needs to have the following properties:

1) Build up viscosity rapidly in AN/H_2O solution at the processing temperature.
2) Property (1) should be achieved with small amounts (1–2%) in the composition.
3) It should be able to take up air when agitated. This is best achieved when rapid agitation is done at a particular viscosity of the mass (not too low or high viscosity).
4) Ability to hold the air/gas bubbles till such time they get fixed in the matrix by cross-linking.
5) Ability to retain the desired micro air bubble structure in long-term storage by not losing its rigid gel structure and not allowing the air bubbles to move and coalesce forming air bubbles of large size whose ability as sensitizer is marginal at best [15, 16].
6) Exhibit a certain amount of pseudoplastic behavior for ease of loading and occupying fully the space in the borehole.
7) Resist syneresis so that bound water does not come out as free water and upset the compositional balance and explosive performance.
8) Resist biological degradation (resistant to bacterial attack).
9) Resist action of salts and metals during the life of the product.

In practice, most of those requirements are capable of being achieved by judiciously choosing the right type of GG, using the most appropriate processing parameters, cross-linking agents, and additives like bubble stabilizers.

A well-formed water gel can have a shelf life of up to 1 year, but it can also be tailor-made to have a lower shelf life if distribution and usage are rapid.

While AN/SN do not interfere with the hydration of GG, another very common ingredient $Ca(NO_3)_2$ causes problems. The calcium ion forms a weak complex with the GG immediately when the gum is added and prevents further viscosity buildup even with prolonged waiting. It has been found that this problem can be overcome by using derivatized gums like hydroxypropyl- or hydroxyethyl-substituted gums or oxidized gums [17]. These build up viscosity in the short term even when Ca ions are present but do not cross-link fast; hence usually depending on the $Ca(NO_3)_2$ content in the OB a mixture of oxidized gum and normal gum can be used. The higher the $Ca(NO_3)_2$ content, greater should be the proportion of oxidized gum. Such a mixture of gums also yields the desired rheology for aeration and stability. For example, at $Ca(NO_3)_2$ content of 9% in the composition ratio of 60/40 oxidized gum to normal gum was found effective.

5.6.2
Specification of Typical GG Used in Water Gels

A typical specification of natural gum useful in many slurries/water gels without more than 3% $Ca(NO_3)_2$ in the composition and capable of accepting air through mechanical agitation is give in Table 5.10.

In the above specification, an important parameter was found to be the protein content. A very low protein content gum was found to be less capable of accepting

Table 5.10 Typical specification of guar gum.

Properties	Specification
Moisture	<10.0%
Ash content	<1.0%
Acid insoluble residue	<5.0%
Protein content	6.0–8.0%
pH of 1.0% aqueous solution	6.0–7.0
Viscosity of 1.0% aqueous solution at 25 °C	
After	
1 h	2500 ± 100 cps
24 h	Not less than 3800 cps
Particle size	
Retention	
on 250 mesh (british standard sieve)	80.0%
on 100 mesh (BSS)	Nil
Cross-linking ability with antimony salts	Must form gels of viscosity >200 000 cps in 24–48 h

air when used in the explosive formation. On the other hand, too high a content led to excessive bubble formation (foaming), which were also rather unstable and the gel underwent degradation rapidly. By experimentation, it was found that a protein content of 6–8% was useful. This protein content can be inherent in the gum. It is also possible to use pure gum (high viscosity) and add to it separately obtained protein-rich materials (usually present in the guar husk (or egg albumin)) to obtain a mixture with the desired protein content and then use in explosive-making process.

5.6.3
Cross-Linking

For slurries cross-linking of the gum have to be of a very low degree in order to maintain flowability of the explosive matrix, at least till it reaches the bottom of the borehole. Additional cross-linker can be added at the last stage if stronger gel is desired in the borehole. Cross-linking phenomenon is essential for making water gels retain their waterproofness, sensitivity, and stability.

This is best achieved by adding cross-linking agent at a stage when viscosity buildup has reached maximum after all the gum has hydrated. The cross-linkers are usually inorganic compounds of antimony, bismuth, and zirconium. Some of them are added as solutions, while some can be added as dry powders mixed with GG. Some of them act rapidly, while some have delayed but prolonged functioning in storage also. Usually a mixture of two types of cross-linker is used to derive maximum benefit. But it has also been found that a single antimony salt (potassium pyroantimonate) can function effectively for most formulations. The concentration of cross-linker to be added will have to be worked out by trial and error depending on the type of gel desired in the end product. Care has to be taken to have the cross-linking just right to avoid instability and syneresis. The percentage of cross-linker in most cases is about 0.04–0.07% of the explosives composition. The quality check is usually the active ingredient content. For example, for potassium pyroantimonate, the antimony content should be close to the theoretical maximum calculated according to molecular formula. Sodium meta-antimonate is another useful cross-linker which can be added as dry powder mixed with the gums. Dichromates, chromates, and borates are also useful as cross-linkers though hot storage stability of dichromate and borate gels is inferior to the gels cross-linked by using antimony salts. pH control is essential for good cross-linking. Retarding the cross-linking action for ease of processing and packaging is also possible. The cross-linking process is time, temperature, and concentration dependent and manipulation of these parameters can give the desired rheology for processability and stability in storage.

The progress of viscosity with time during processing after addition of cross-linker for some combination of GGs and starch is given in Table 5.11.

Instability of gel structure in the explosive is brought about by

1) AL/H_2O reaction
2) Biodegradation of gums

Table 5.11 Viscosity of matrix blend as function of time for various thickeners.

Composition	Viscosity X1000 cps				
	Elapsed time in minutes				
	10	30	50	120	24 × 60
AN/SN/CN/guar gum	21	23	25	25	25
AN/SN/CN/guar gum/oxidized gum	55	—	65	90	>200
AN/SN/CN/oxidized gum/tapioca starch	55	65	70	85	>200
AN/SN/CN/guar gum/tapioca starch	40	50	—	65	65

3) Oxidation
4) Over-cross-linking
5) Partial hydration of gum
6) Incorrect pH
7) Gum not according to specification
8) It is not GG at all
9) Poor dispersion of gum.

5.6.4
Mechanism of Hydration

Gums other than guar are also capable of hydrating and cross-linking, but so far GG has proved to be the one most used for ease of availability, cost, and desirable properties. Other gums known to have been tried out are locust bean gum, gum karaya, gum arabic, gum acacia, xanthan gum, and psyllium husk. In the author's experience, xanthan gum is more than capable of replacing GG but is more expensive and not available everywhere. Gum acacia is used more as a bubble stabilizer. Author's experience has been positive in this respect and gum acacia imparted additional sensitivity over prolonged storage in explosives. This was particularly evident in nonaluminized formulas.

GG belongs to a class of natural product family name Leguminosae that can be dispersed in H_2O to obtain a viscous solution. It is found as a seed in the pods of the guar plant in India and Pakistan. The spherical brownish seeds are flamed, loose skin removed, endosperm is separated from the germ by milling, and the so-called splits are known as *guar flour*. The guar flour is refined further by processing including precipitation by ethyl alcohol, filtered, dried, sieved, blended into a whitish powder, bland in taste, and edible. The refined guar is classified into different grades based on particle size, dispersion ability, rate of hydration, and viscosity leaving the choice to the end user in order to suit his requirements [18].

The GG is a water-soluble polysaccharide and a galactomannan. Required water-soluble polysaccharide content, ideally is 86% out of which 36.6% is

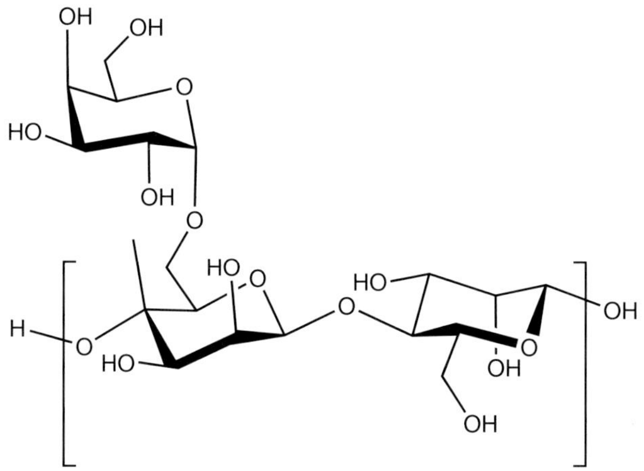

Figure 5.3 Structure of Guar Molecule

D-galactose anhydride and 63.1% is mannose. In practice depending on the plant origin, this ratio can vary. The chemical structure of guar has been extensively studied using chemical and instrumental analysis and the conclusions as to the actual structure are in good agreement. The structure most accepted is in p. 47 of [18]. The molecular weight is around 220 000 and it is a straight chain galactomannan with galactose on every other mannose unit. Beta 1–4 glycoside linkages couple the mannose units and the galactose side chains are linked through alpha 1–6 linkages. The mannose to galactose ratio is approximately 1.9 to 1.0. The richer the galactose content, the more soluble is the GG in water. Guar has the property of dissolving with ease in cold water attaining its full viscosity in due course. The GG shows great affinity to water molecules and according to well-accepted postulations, this is achieved through hydrogen bonding. GG as such can also be modified by reacting its hydroxyl groups by esterification, ethoxylation (reaction with ethylene/propylene oxide) complexing, and cross-linking where the linear chains are drawn into a three-dimensional network (Figure 5.3). Varying degrees of polymerization or modification allow control of hydrolysis. Viscosities are measured by Brookfield viscometer or by Ford cup; hence the values may be termed as *apparent viscosity*. The viscosity depends on shear rate. The viscosity measured also depends on efficient dispersion. If not properly dispersed, agglomerates hydrate and prevent gum particles inside from coming in contact with water and thus full viscosity is not realized. In order to disperse better, wetting of guar powder with retardants like glycerine, ethylene glycol, or mixing with other dry ingredients like sugar can be adopted. Viscosity is temperature and concentration dependant. The unique properties of GG very useful in industrial application are its ability to build up viscosity even in cold, retain its viscosity even at high temperature (80 °C) for considerable length of time, good solubility over wide pH range(4–10.5), resistance to electrolytes being nonionic in character (unaffected by hard water).

Table 5.12 Viscosity of guar gum at different temperatures and time.

Viscosity of 1.0% aqueous solution	At 25 °C 4000 cps after 5 h	At 85 °C 6000 cps after 10 min
Viscosity of 1.0% aqueous solution	At 27 °C	At 27 °C
Good guar gum	2700 cps after 1 h	3800 cps after 24 h
Poor quality guar gum	2400 cps after 1 h	No viscosity
		Solution disintegrated

GG is also very compatible with other thickeners and viscosity builders like starch, and this is made use of in building specific combinations for specific requirements. The hydration ability of guar gets affected only in the presence of substances which compete for bonding with the water molecules such as salts with water of crystallization like $Ca(NO_3)_2$. Table 5.12 shows the variations in hydration.

5.7
Permissible Explosives

5.7.1
Design Criteria

Water gel explosives especially designed for safe use in underground coal mines [1, 4, 10] were the first non-NG explosives to be produced and used in large quantities. Today in most countries, either water gels or emulsions are used exclusively for underground blasting in gassy coal mines.

Underground coal mining has been historically resorted to by humans from ancient times. The availability of very good quality of coal (high calorific value/low ash content) underground was exploited in countries like Europe (UK, France, Germany, Poland, etc.), USA, India, South Africa, and China to a great extent. Till about 30 years ago in most countries except USA/Australia, the percentage of coal mined from underground exceeded that from open-cast surface mines. Today underground mining of coal is still pursued even though not increasing in volumes as rapidly as open-cast mining due to difficult mining conditions and economic reasons which can vary from country to country. In most countries production of coal from open-cast mining dominates the scene, but in some other countries like India the quality of coal obtained underground is superior and hence underground mining with explosives is still being pursued.

For purposes of functioning as a permissible explosive, the following needs to be achieved in order to qualify for use in underground coal mines:

1) Be safe/have low risk of igniting methane/air mixture which could be formed by emission of methane in underground gassy coal mines.
2) Be safe against deflagration, that is, prolonged burning of explosives instead of fast detonation which increases risks of methane/air catching fire and exploding.

Table 5.13   Range of ingredients in permissible water gels.

Ingredients	Quantity (%)
Ammonium nitrate	45–60
Sodium nitrate	8–10
Calcium nitrate	6–8
Sodium chloride	5–10
Water	15–25
Guar gum	1.0–2.0
Cross-linker	0.04–0.06
Aluminum flake powder	3.0–3.5
Sugar	6.0–8.0

3) Should be safe and have low propensity to set off coal dust/air mixture formed during blasting operations.
4) Should not get pressure desensitized due to channel effect or from another shock wave coming from a nearby explosive which has denoted already due to delay detonators being used for more productive mining.
5) Should be able to jump air gap and preserve COD in small diameters as coal particles can intercede between two cartridges in the borehole.
6) Should be cap-sensitive to permitted detonators (due to vibrations and noise, the use of detonating fuse and booster is avoided for priming).
7) Should have reasonable strength and gas volumes to enable economic working of the mine by blasting.
8) Should have minimum toxic fumes generation.
9) Should have resistance to action of water.

5.7.2
Tests for Permissibility

The tests which are imposed to qualify as a permissible explosive can differ from country to country depending on the method of blasting, presence of methane, and ventilation facilities. The tests followed in USA are different from that standardized in UK. India follows the tests similar to those in UK. Germany and France have different testing methods.

But all of them are methods to gauge the risk of the explosive setting off a methane/air or coal/air explosion when functioning normally in blasting.

Generally in the so-called gallery tests, the explosives are placed in the borehole of a cannon whose business end projects the shock wave and the hot gases into a long steel cylinder called the *gallery* which can have methane/air mixture or coal dust suspended in air as required at the time of initiation of the explosives under test. These are created artificially to stimulate actual field conditions. It has been found by experiments that 9% pure methane with air is the most sensitive mixture for ignition. Natural gas at 8% in air has the same tendency to ignite as 9% pure methane air mixture because of its composition containing other hydrocarbons which are more active. The explosives to be tested are loaded in the cannon and fired. Ignition or no ignition is recorded by observing the opposite end of the gallery. In an ignition, flame and hot gases are thrown out from the gallery end (opposite side to the cannon end) and can be easily observed visually. The point of initiation of the column of explosive in the cannon and whether stemming is used or not are all factors influencing the test results and based on experience the test procedures have been standardized. Details of testing done in UK/France/USA/India are given in Appendix A.

5.7.3
Other Tests requirement

Toxic gases produced postblast, air gap sensitivity, COD, and cap-sensitivity are other properties evaluated. The pass/fail criteria for these tests are also different from country to country.

5.7.3.1   **Deflagration Tests**
Deflagration in explosives is burning rather than detonation. Low explosives such as black powder deflagrate in the open almost all the time and burn till they are consumed. Only under confinement they tend to go over to detonation.

Deflagration is dangerous in coal mines where air/methane mixtures could be present. Prolonged burning of explosive cartridge if it happens in total confinement is not so dangerous as when it occurs with the explosive exposed to atmosphere. Deflagration occurs when the explosive receives weak initiation or is in a blown out shot or due to pressure desensitization. In a blown out shot after initiation if the subsequent cartridges with the explosive in contact are borderline-sensitive and fail to move the coal then instead of detonating, they will deflagrate or remain without reaction. The latter is much more preferable from the point of view of safety. Thus it is important to evaluate the deflagration tendency of an explosive and tests are available for this purpose. Although these differ from country to country, the basic principle is to expose the explosive to a weak detonation wave which can be from burning coal dust/black powder and study the behavior of the explosive being tested. If the explosive being tested also burns, it is noted as having a deflagrating tendency and if found intact it is termed as *nondeflagrating*.

The NG explosives were designed and thoroughly tested for deflagration, and only specially designed formulations could pass the stringent measures of the test. The explosives passing the tests were found very weak in strength. The permissible water gel explosives have been tested extensively in the 1970–1980s and have shown in general less tendency to deflagrate than NG-based explosives. Surprisingly even the water gels containing AL flake powders as sensitizers were able to pass the deflagration tests.

The Audibert Delmas test method has been used extensively and considered a valid test for evaluating deflagration risk in an explosive.

In this method a test cartridge is placed coaxially in a steel tube and the annular space as well as the top is filled with coal dust (standardized as to its properties) and tamped. A deflagrating low explosive is placed on top of the test cartridge and a cap with an orifice (2–3 mm) fitted. The deflagrating explosive is initiated and the behavior of test cartridge is noted as pass/fail depending on whether it deflagrates or not.

Cerchar's Method (France) Here in a hermetically sealed tube, the cartridge comes under the influence of burning black powder whose mass is varied in steps in each test. It is assumed that probability of deflagration increases with the mass of the contributing material (black powder) and by the usual up and down trials 50% probability point called the *deflagration mass limit* is determined. Evaluation can then be made by comparison of mass limit values.

USBM Test for Deflagration This is similar to Audibert Delmas test except coal dust is standardized to 37% volatiles. The end of pipe has a rupture disk set at about 600 psi. Detonator is used as donor charge and is positioned at varying distances from the cartridge. If the explosive is cap-sensitive at short distances from the detonator, it should detonate or fail (no reaction) at large distances. Intermediate distance initiation could lead to deflagration because of weak initiation. The distance criteria for 50% probability of deflagration could be used for risk evaluation.

5.7.4
Behavior of Water Gels in Permissibility Tests

Historically it was a revolutionary discovery that water-based explosives could be made cap-sensitive without addition of self-explosive ingredients. More astonishing was the work of USBM [4, 5] that revealed that water gels containing AL (paint-grade) flake powder were capable of not only being cap-sensitive but also of passing permissibility tests when formulated in a particular manner. It was demonstrated by them that water gel explosives containing H_2O, NaCl, and paint-grade Al powder could pass US gallery tests comfortably. It was left to Gehrig and Mahadevan to extend their works and formulations which could pass testing of permissibility as laid down by SMRE Buxton, UK, CMRS India, and Dortmund-Derne Germany were established. It was also demonstrated that these explosives also showed greater disinclination to deflagrate compared to similar category permissible explosives based on NG/nitrocotton. Some of these water gel formulations contained coolants (NaCl, H_2O), flame suppressors (oxalates, phosphates), and even the presence of perchlorates as additional oxidizer did not bring down the ability to withstand deflagration. The BAM Koenen test was used for those experiments. The results were astounding as quite a high percentage (6%) of Al was used in the composition and presence of AL was considered as taboo in underground mining. It was with great difficulty and with backing of considerable experimental factual data that permissible water gels were allowed as commercial product. One needs to understand the apprehensions of the mines safety authorities in this respect as after nearly 100 years of using NG explosives in underground gassy mines, a new type of explosive had appeared with metal powder (AL) as one of its ingredients. Ignition of methane/air mixture in underground mines while blasting is thought possible due to

1) direct action of shock wave from the explosive;
2) indirect action of shock wave after reflection from solid surfaces surrounding the explosive column;
3) effect of hot gases produced in detonation;
4) effect of hot solid particles from detonating explosives; and
5) prolonged burning of sold residues after detonation.

Reasons (3) and (4) eventually can be related to explosion temperature attained during detonation of the explosives as also the duration of the flame.

It was somewhat puzzling to note that even the higher VOD of the water gel explosives did not have a negative effect on the permissibility. The VODs measured in the open, in 32 mm (11/4 in.) diameter cartridges of water gel explosive in good condition consistently were between 3500 and 4000 m/s, at least 10–15% higher than permitted NG-based explosives. However the flame temperature measured was around 1500 °C, lower than that in NG explosives. Duration of the flame could not be consistently measured for lack of proper instrumentation at the time of interest. It appears that influence of shock wave on permissibility is less than that of flame temperature. Even the emulsion explosives which exhibit even higher VODs (4500–4800 m/s) have also passed permissibility tests. The after burning of metallic particles such as AL needs to be fully avoided to circumvent the effect of burning solid particles igniting the methane/air mixture or coal dust atmosphere. The fine paint-grade Al powder has not proved a threat to safety at all. More than 500 000 tons of aluminized water gels have been consumed without mishap in the underground coal mines classified as requiring permissible explosives for blasting. One cannot also ignore the role played by water in the formulations of permissible products. The water content ranges from 15% to 25%; higher the water content, higher is the ability to pass the permissibility tests provided the explosive is made sensitive enough to detonate fully when initiated with a No. 6 detonator. Water acts as a heat quencher but produces higher gas volume in detonation.

Pressure desensitization and deflagration were the other two factors which had to be addressed when the water gel explosives were commercially introduced in underground coal mining. Both these were not experienced in field tests. It was thought that sensitization through air bubbles in contact with energetic metal present as a thin flake in a soft compressible matrix with spring back properties enabled the explosive to withstand pressure desensitization to such an extent that it was practicable to use them in blasting in coal with delay intervals between the boreholes [19]. On the other hand, channel effect could influence the product performance negatively but could be overcome by resorting to tamping to fill the borehole completely with no annular free space for the shock wave to advance ahead of the decomposition through air/rock medium and increase the density leading to higher insensitivity of the explosives awaiting detonation. When microballoons were used, if walls were thick, then pressure desensitization was noticed more often. Based on this feedback, manufactures of hollow microspheres have made microbubbles with very thin walls (but of sufficient strength for handling without disintegration) so that the air bubbles are not prone to permanent desensitization by passage of detonation front. Some amount of elasticity has also been inbuilt by using some special ingredients (polymers) as material for the hollow balloons. Air-gap sensitivity was also found to be more than adequate (>3 in.) if proper quality of paint-grade AL powder was used. It is interesting to note that additional AL in the form of atomized powder while increasing critical density of the product and its strength brought down substantially the permissibility rating. In fact such formulations containing atomized powder in low percentages of 1–3% when made into explosives were barely able to pass the test stipulated for the lowest permissible category rating. The plausible explanation is the negative influence of atomized particles burning with intense heat and cooling off much slower than paint-grade AL powder, thereby increasing the chances of igniting the methane/air mixture.

Water gels are known to exhibit air-gap sensitivity of 2 in. and above (even up to 6 in.) in 1-in.-diameter cartridges and the gap sensitivity increases with charge diameter and confinement both for permissible and nonpermissible types. Water and NaCl have proved beneficial in maintaining explosive properties as they help in holding the AN in solution. Surprisingly at constant H_2O content, addition of NaCl increased VOD and gap sensitivity for compositions where all the NaCl was fully dissolved in H_2O (5% NaCl for H_2O content of 18%). When NaCl is present in crystalline/granular form as a solid, it once again reduced explosive properties but permissibility increased due to higher NaCl content bringing in additional cooling effect to act against incendivity. The role played by NaCl and water in making the water gel explosive pass the permissibility tests with comparative ease has not been fully studied. It would be indeed worthwhile to conduct deeper studies in order to understand the mechanisms involved, although in practice formulations have already been arrived at by trial and error and from historical knowledge, capable of passing the most severe tests for permissibility.

Air bubble sensitization [15] is important in permissible explosives as much as in general purpose product. As described earlier, in order to maintain sensitivity (even under pressure) artificially encapsulated air bubbles have been tried out successfully. However economic considerations have prevented large-scale usage. The use of such microballoons ensures uniform presence of the microstructure of greater permanence than mechanically or chemically introduced air/gas bubbles. The only disadvantage could be in the process of addition and dispersion because of high bulk volume as also the probable increase in hazard of the explosive containing a frictionable material while being pumped or packed.

Air bubbles introduced through hollow glass spheres by themselves, that is, in compositions without AL or MMAN as sensitizers, have not been able to impart cap-sensitivity in water gels/slurries even if the density is taken below 1.00 g/cc.

5.7.5
Toxic Fumes and Typical Formulation

Underground coal mines require explosives with minimum production of toxic fumes in order to have safe working conditions but adequate gas volume is necessary for efficient breaking of coal.

The earlier NG explosives were not good in this respect and approach to working face was delayed considerably while waiting for the toxic fumes to be cleared. Water gels can be oxygen balanced with ease and will have less toxic gases postdetonation. Toxic gases volumes are much less than limits set (less than 2.5 ft^3/b of explosive). However the water gel must be in good condition and detonate fully. A typical permissible water gel could have the following basic composition range [Table 5.13]:

Oxygen balance is adjusted for the composition to 0.0 to -3.0.

When polythene or paper is used for packing, the composition will have negative OB to a greater extent.

Other ingredients which can be used are Na perchlorate, calcium nitrate, ammonium chloride, oxalates, sugar, and urea each fulfilling the role of oxygen supplier/fuel/coolant. Many other ingredients have been claimed in patents to contribute to the overall performance/permissibility of water gels but most of the formulas of commercially produced permissible products are built around the ingredients discussed above and in (Table 5.13). The technology of production could vary.

5.7.6
Strength of Permissible Water Gels

Within the statutory requirements of permissibility-based tests, water gels are able to compete quite well with NG explosion on strength and performance in actual blasting. Although having density of 1.10–1.20 g/cc as against gelatinous permissible water gels (having density of 1.25–1.40 g/cc), because of their higher gas volume and VOD they are able to compensate for the low density. The overall strength difference between the two is marginal. On the other hand, the water gels perform better compared to NG-based powder explosives or semigelatins whose density ranges from 0.98 to 1.20 g/cc. The added safety which the use of water gels brings in blown out shots (no deflagration), while drilling relief holes, less toxic fumes, and no headaches due to physical contact have made it preferable for underground usage especially in coal mining.

In the rifle bullet tests, explosives were kept at 30 m distance on an mild steel backing plate and a bullet fired from a pill box using a 0.303 rifle. Whereas the NG explosive could be initiated, the permissible water gel failed to initiate. In the drilling test, a remote-controlled drill rod was used to drill through a cartridge kept on a bed of a coal and rubble. The NG-based explosive detonated soon after the drill bit penetrated the cartridge, whereas in case of the water gel there was no detonation even after the drill bit pushed through the explosive completely. Regarding generation of toxic fumes, in one study at the same mine face underground after blasting, face could be approached only after 1 h of waiting in case of NG-based explosive, whereas they could approach in 20 min when water gel was used. Thus considerable productivity increase could be expected by use of water gels.

5.8
General Purpose Small-Diameter Explosives (GPSD)

While water gels made inroads into underground coal mining after initial resistance from mine safety regulators due to the presence of AL/perchlorate, it was a greater challenge for the water gels to succeed in the areas of tunneling (noncoal) in hard strata, in quarrying of stone and in deepening of wells, which application is very prevalent in countries of Asia and Africa, from the point of performance.

5.8.1
Design Criteria and Composition

The criteria for a preferred explosive in these areas of application are primarily

- strength (weight and volume),
- density and toxic fume generation,
- long shelf life,
- cold temperature sensitivity,
- stability in hot storage, and
- safety in handling and storage.

In the safety requirements water gels/emulsions could show significant improvement in potential risk due to their lower impact/friction sensitivity as compared to NG-based explosives. However it has taken considerable time, research, and development to satisfy other requirements of the end user who could be large corporate mining companies, blasting contractors, and individual farmers.

The weight and volume strengths difference between water gels at 1.25 g/cc as against 1.40 of gelatins have been bridged to a great extent by higher gas volumes and by using highly energetic ingredients in the compositions to compensate for the absence of NG. Experimentally obtained values are given in Table 5.14.

The design of general purpose small-diameter (GPSD) is somewhat easier than a permissible product as there are no stringent test criteria. A greater choice of oxidizer/fuels/sensitizers/strength enhances are available to the formulator. Water content can also be lowered as compared to permissibles thereby pushing up strength and density but also increasing sensitivity to friction impact, fire, and static. Some of the GPSD tested answered positively to rifle bullet test. Typical formula for GPSD is given in Table 5.15.

In case an organic sensitizer like MMAN [3] is used, then flake powder could be reduced or omitted. The strength of the GPSD explosive could be increased by adding more of sodium perchlorate, atomized AL, and sulfur. The latter, however, increases the toxic fumes by liberating SO_2 in the reaction with $NaNO_3$ although

Table 5.14   Explosive properties of NG-based explosive versus water gels.

Properties	Special gelatin 80%	Water gel
Density (g/cc)	1.40	1.25
VOD in open (25 mm diameter)	3500	4000
Total gas volume (l/kg)	750	825
Lead block expansion (ml)	340	325

Table 5.15   Typical GPSD formulation range.

Ingredients	Range (%)
Ammonium nitrate	60–77
Sodium nitrate	5–8
Water	8–12
Sodium perchlorate	4–6
Gums	0.8–1.2
Fuel	4–6
Aluminum atomized powder	3–5
Aluminum flake powder	3–4
Sulfur	2–4
Ammonium nitrate porous prills	5–10

the total gas produced increases due to liberation of both SO_2 and N_2.

$$6NaNO_3 + 6S = 3Na_2SO_4 + 3N_2 + 3SO_2.$$

5.9
Sensitizers

5.9.1
Inorganic

Addition of AN porous prills added at the end of the mixing cycle termed *"doping"* is used to increase the AN content without increasing the processing temperatures or creating a very crystalline product with lower sensitivity. In fact some manufacturers keep almost 30% of the AN out of the oxidizer solution so that crystallization point is lowered and processing can be done at a lower temperature. There is a question about the AN added on top fully participating and contributing to the total energy as being only mixed later they may not be in intimate contact with fuel at the time detonation is taking place in the explosive. It is possible that AN is vaporized and decomposed rather than detonated and it may merely fulfill the role of an oxygen supplier which can be made use of only if required fuel is also present in contact with it. One way out of this would be to add AN/FO of 94/6 variety (fully oxygen balanced) instead of AN alone, but the FO desensitizes AL powder, affects the surface tension of water/AL interface, and also disturbs the stearic acid coating which enables air bubble to remain in contact with AL flake particle. Benefit of doping can be derived only if highly porous sensitive prills are added. Of course this also lowers the density of the product with usual consequence of lowering the energy available in a fixed volume of the borehole. Chemical gassing and generation of air bubbles inside the matrix after the material is packed could ensure continued sensitivity even during longer storage in formulations with high AN content. It is also possible to modify the AN crystals coming out from the supersaturated solution in such a way by addition of crystal habit modifiers that they are more porous and hence more sensitive. A high degree of control is necessary to prevent overdoing this effect as again it will lead to lowered density, whereas the aim in GPSD is always to increase the density of the product. This is best achieved if starting density of the OB is kept as high as possible, but a limit is reached in respect to lowest water content practically feasible. Sodium perchlorate and $Ca(NO_3)_2$ contribute to increase OB density but both are available with inbuilt water, one as aqueous solution with 30% water and the other with water of crystallization and hence a limit is reached here also.

Density of OB to make a GPSD with AN 55%, SN 8%, CN 9%, SPC 6%, and water 10% is 1.60 g/cc, whereas density of an OB to give a GPSD with AN 64%, SN 8%, and water 12% is only 1.42 g/cc. A 20% air inclusion would be sufficient for getting the desired sensitivity if the bubbles are of the right size. This would bring down the density in the $AN/SN/SPC/Ca(NO_3)_2$ from 1.6 to 1.3 g/cc region, whereas the $AN/SN/H_2O$ would come down from 1.42 to 1.14 g/cc, a density far lower than 1.45 g/cc of an NG-based GPSD explosive. As OB is 86–88% of the explosive composition, it is worthwhile to keep its density as high as practical although from processing/safety considerations a limit will be reached especially for the water content of the explosive.

5.9.2
Organic Sensitizers

Organic (nitrate) sensitizers have been tried out as sensitizers in water gels instead of AL flake powder whose cost and handling posed some difficulties. It was also thought that organic nitrates could be used in larger quantities without creating a stiff material difficult to pack in a continuous chub machine. The organic sensitizers tried out are MMAN, ethylenediaminedinitrate, ethylenediaminedinitrate, ethanolamine nitrate, and nitromethane. The most used organic sensitizer is MMAN and the DuPont Co (USA, Canada) was the first to commercially offer a series of water gel explosives based on MMAN under the brand name TOVEX. Canada, South Africa, and India also produce these types of explosives. MMAN is highly soluble in water and can be easily made by absorbing MA in

HNO_3 acid. Usually MMAN is handled as 86% solution. At this concentration, it is considered safe to handle. MA is available as a byproduct in the petrochemical industry. In order to produce more concentrated MMAN, MA gas itself is absorbed in concentrated nitric acid. MMAN can also be chemically formed when AN is made to react with HCHO [20]. The reaction products are mono-, di-, and trimethylamine nitrates, but MMAN is the major component. The reaction mixture containing water can be directly used to make the explosive, the only disadvantage being that di- and trimethylamine nitrates contribute little to the explosive energy and also disturb the gel structure when GG is used as a thickener.

It has been noted that MMAN is prone to mass effect even in solution and an accident which occurred in 1974 in a loosely shunted rail tanker in the USA alerted users of the possible risk hazard. Author's own experience in the laboratory where several ignitions took place during preparation of MMAN showed that as water content is reduced from 14%, the hazard of spontaneous ignition and possible explosion increases dramatically. At about 6% H_2O, the risk becomes quite unacceptable as even slight heating results in fire. Nevertheless at greater dilutions MMAN is being used in making water gels, even up to 30% in the composition imparting cap-sensitivity and high explosive strength to the product. When converted into small-diameter explosives, presence of micro air bubbles was necessary to sustain initiation and detonation at MMAN concentration of 20–30% along with AN/Ca $(NO_3)_2$/SN/per chlorate. Dry mix of these oxidizer salts and MMAN are quite sensitive to impact/friction to a greater degree than MMAN alone, and where pumping MMAN/explosive is resorted to lot of care needs to be taken to avoid dry running. MMAN crystals are also quite sensitive, especially when mixed with hard gritty particles.

5.9.3
Air/Gas/Synthetic Bubble Sensitizers

Air bubble incorporation is used in water gels with AL flake/MMAN powder as additional insurance against loss of sensitivity. We have already discussed mechanical aeration through high-speed mixing and through addition of AL flake powder of low bulk density. Another method of aeration can be through chemical gassing where chemicals which decompose in the presence of explosive matrix at certain pH values release nitrogen gas bubbles, which are entrapped in the gelation process and remain in the explosive matrix throughout the life of the explosive. It is necessary to add the right amount of gassing agent at the right time in the process. The release of gas is due to chemical reaction which is temperature-, pH-, and time-dependent. All these parameters need to be manipulated to get the desired result of having microsized bubbles evenly distributed throughout the explosive matrix and of optimum number (neither too low nor too large) to affect density values out of desired range. Spherical droplets are the best in regard to stability [16] and are achieved only when they are introduced in a media of suitable rigidity (viscosity/cross-linking). While the process of gassing is simple to control,

Table 5.16 Time of high speed mixing versus density.

Time of mixing (min)	Density (g/cc)	Temperature of mix ($^\circ$C)
0	1.38	60
5	1.30	54
10	1.22	49
15	1.15	45
20	1.16	43
25	1.20	42
30	1.26	40

it effectively needs much experimentation in laboratory and pilot plant batches before it can be established in production process. Several gassing systems are known and practiced both for sensitizing slurry/water gel explosives and emulsion explosives as well. Details regarding these systems are presented separately. The most commonly used system is sodium nitrite coupled with sodium thiocyanate or with urea in acetic acid medium.

If one resorts to use of man-made bubbles (hollow microspheres), then control is much easier as these are uniform and require only to be distributed evenly in the matrix to give consistent firing results. Such microspheres are manufactured commercially only by a few. Although costs have been brought down substantially in the last 5 years still they are expensive and form a large part of the RM cost. Their high bulk also presents handling difficulties similar to AL powder though not having the same hazard risk. Microspheres are available with thick- and thin-walled construction and in different particle sizes. Thin-walled, finer product introduces greater sensitivity to the explosive mix.

It was also noticed in earlier years that presence of microballoons in SD explosives when used in underground mining could lead to pressure desensitization of explosive leading to poor performance. It was thought that passage of shock wave through the explosive broke up the bubbles irreversibly before hotspots could be formed and the density also went up beyond the critical density at that diameter. However recent claims from manufacturers of the microballoons indicate that using thinner walls, thermoplastic material has overcome the problem of shock desensitization. Probably due to some amount of elasticity in their structure, these hollow spheres are able to function as locations of hotspots, thereby helping propagation of the detonation in the explosive even after the passage of a shock wave from the neighboring explosion through it.

Generally there are three varieties of microballoons:

- Glass microspheres (from 3 M Corp., USA)
- Thermoplastic spheres (Expancel from Akzo Nobel, Sweden)
- Silicon-based (from PQ Corp., Australia).

These microspheres have been particularly recommended for usage in emulsion explosives where AL flake powder is nonfunctional as a sensitizer. Experimental results show slight improvement in VOD, higher critical density, and most of all consistent behavior under hydrostatic pressure and rock burden. If economics permit, inclusion of small percentage of microballoons in addition to mechanical aeration or chemical gassing would be the best way of utilizing the advantages of microballoons without going overboard on costs. Once again problems associated with even dispersion in explosive matrix have been taken care of by the manufactures by prewetting the product with liquid which is compatible with the explosive gel.

When bubbles are incorporated into the matrix by mechanical agitation or by gassing, they tend to rise and escape [15, 16]. If not prevented by a rigid matrix, they tend to migrate from higher pressure area (smaller size bubbles) to lower pressure area (bigger bubbles). Thus the larger bubbles tend to grow and smaller ones shrink in number and hence explosives' sensitivity gets affected. Use of surface-active agents in lowering the surface tension is extremely helpful in incorporation of air during stirring, but if AL flake powders are used the effect of these surface-active chemicals on AL flake powders surface hydrophobic properties needs to be watched. For example, sodium lauryl sulfate (SLS) in very small percentages had no effect but any quantity more than 0.1% desensitized the explosive on storage.

Micrograph studies or even observation using laboratory microscope has shown that bubbles under 40 μm were very effective in sensitizing the explosive. While obtaining these size bubbles is not a problem, the entire life of the explosive gel is dependent on keeping them at that size. The longer this can be done, the longer is the useful (shelf) life of the explosive at its maximum potential. Naturally much depends on the rigidly and stability of the cross-linked gel. Any loosening of the gel structure will enable the gas bubble to grow and escape. Instances are there when the top of the cartridge contains all the air put in, while the explosives now at high density sit at the bottom in a separated fluid state. pH, GG concentration, degree of hydration, and cross-linking are the critical factors in providing a gel matrix for a stable, long bubble life of the desired size. It is also pertinent to note that an explosive even while preserving its overall density becomes less sensitive and VOD drops. This is attributed to the bubbles growing in size and becoming less effective as hotspots (the heat produced by adiabatic compression is less than the energy absorbed). A typical example is of NG which becomes extremely sensitive if microsized bubbles are present, but not so if large pockets of air are occluded. The importance of maintaining pH in keeping the gel structure in the best condition necessitates the use of buffers to prevent wandering of the pH from the desired range of 4.5–6.0. Use of lignosulfonate is also beneficial in decreasing bubble size. Surface-active properties influence rate of aeration, bubble size distribution, and gas permeability in stirred aqueous gels.

The mechanical aeration of a viscous solution of oxidizer salts in H_2O thickened by means of GG is perhaps the easiest way of sensitizing explosives, but the density control is dependent on time of mixing which is *inter alia* dependant on viscosity and temperature. Effective aeration attains a peak and then further mixing does not improve either in terms of volume of air or size of the bubble (Table 5.16). On the other hand, high-speed agitation prolonged beyond a certain time drives out the air and also increases the size of the bubble. Even the gel quality gets negatively influenced by overmixing at high speeds as it leads to less elastic gel, a key to long-term stability.

Use of water-soluble polymers like polyacrylamide and cross-linking it instead of a guar system has also been tried out and found successful. But not enough information is available as to its use on a regular basis commercially, perhaps more for economic reasons than technical. A good guar-based gel system has proved economical and sufficient for most applications. The polymer system may be useful where organic sensitizers are used as the presence of organic sensitizer could affect the stability of the guar gel or even hydration of the GG.

References

1. Cook, M.A. (1958) *Science of High Explosives*, ACS monograph 139, Rheinhold, USA
2. Mahadevan, E.G. (1976) Recent Developments in Technology of Slurry Explosives. Int. Jahrestag, Inst. Chem. Treib, CA89 (4)26968J, pp. 15–29.
3. Robinson, R.V. (1969) Water gel Explosives-three generations–Canadian Mining and metallurgical Bulletin. *Can. Min. Met.*, 1317–1325.
4. Van Dolah, R.W., Mason, C.M., and Forshey, D.R. (1968) Development of Slurry Explosives for Use in Inflammable Gas Atmospheres. Report No. 7195, USBM.
5. Hay, J.E., Watson, R.W., and Van Dolah, R.W. (1973) Development of watergel permissible explosives. 15th International Conference on Safety in Mines, Karlovy, Czechoslovakia.
6. Urbanski, T. (1976) *Chemistry and Technology of Explosives*, vol. IV, Pergamon Press, New York, pp. 546–553.
7. Alcoa (1975) Alcoa Aluminite Powders in Blasting Agents, Section FA2D-6, Powders and Pigments.
8. Edwards, J.D. and Wray, R.I. (1955) *Aluminum Paint and Powder*, Rheinhold Publishing Corp., New York.
9. Aluminum and Health (1984) Aluminium and Health, Aluminium and Health, The Aluminum Association, Inc., Washington, DC.
10. Mahadevan, E.G. (1971) Influence of tropical storage conditions on stability and performance of High explosives especially Permissible Slurries. ICT Annual Seminar, Karlsruhe, Germany.
11. Goring, D.A.I. and Young, E.G. (1955) *Can. J. Chem.*, **3**, 480.
12. *Recommendations for Storage and Handling of Aluminum Powders and Paste* TR-2, (1957) Aluminum Association Inc., Washington, DC.
13. Whistler, R.L. and Bemiker, J.N. (1973) *Industrial Gums*, Academic Press, New York.
14. Whistler, R.L. (1954) *Natural Plant Hydrocolloids*, Advances in Chemistry Series, Vol. **11**, American Chemical Society, Washington, DC
15. Keirstead, K. and DeKee, D. (1980) Stable bubble sensitized gel slurry explosives. *Ind. Eng. Chem. Prod. Res. Dev.*, **19** (11), 91–97.
16. Kumar, R. and Kuloor, N.R. (1970) The formation of bubbles and drops. *Adv. Chem. Eng.*, **8**, 256.
17. Subbiah, G., Sethuraman, B., Mahadevan, E.G., and Rao, T.N. (1978) Kin. methylation of primary alkyl amine hydrochlorides with formaldehyde. *J. Indian Chem. Sect B.*, **168**, pp 1009–11.
18. Chudgikowski, R.J. (1971) Guar gum and its applications. *J. Soc. Cosmet. Chem.*, **22**, 43–60.
19. Wieland, M. (1986) The dynamic desensitization of coal mining explosives. 13th Symposium Franklin Research Center on Explosives and Pyrotechnics, Philadelphia, PA.

20. Venkatiah, S. and Mahadevan, E.G. (1982) Rheological properties of hydroxypropyl and sodium carboxymethyl substituted guar gums in aqueous solutions. *J. Appl. Polym. Sci.*, **27** (5), 1533–1548.

Mahadevan, E.G. and Urbanski, T. (1974) Water-gel (slurry) Explosives. Lecture notes, India.

Mahadevan, E.G. (1981) Watergel explosives. *J. Chem. Stosow.*, **25** (3), 345–367.

Further Reading

Carlsson, W., AZieglenfus, E.M., and Overton, J.D. (1962) Compatibility and manipulation of guar gum. *Food Technol.*, **16**, 50–54.

6
Emulsion Explosives

6.1
Introduction

The issues of sensitivity and performance in small-diameter blasting, water proofness, and dependence on the quality of a single raw material (RM) namely AN prills caused enough concern to usher in products other than ammonium nitrate/fuel oil (AN/FO). Thus, it was logical to build up other systems of oxidizer/fuel/energizer/sensitizer to arrive at new types of explosives. Research and development yielded two systems (i) slurries/water gels and (ii) emulsions. We have already discussed in detail the technology of slurries/water gels thoroughly in Chapter 5. We shall now take up the study of emulsion explosives in detail.

Historically, the USA is credited as having first spawned the emulsion explosive [1] in 1968 although claims of knowledge of such type of explosive existed from other countries (Sweden, China, and Canada). The first commercial quantities of emulsion explosives were produced in India based on technology from ATLAS Chemical Industries, USA, in 1971. The product made in India was a large-diameter cartridged product with booster sensitivity. Few years later pumpable emulsions and still later small-diameter cap-sensitive products also became available to the mining industry. Initially the technology, formulations used produced emulsion explosives with some problems of storage stability and sensitivity as also process difficulties and safety issues. However, in the last 15 years these problems have been sorted out one way or the other and today the technology has matured and established itself as a very viable concept of making emulsion explosives to cater to most of the end user requirements [2].

6.2
Concept of Emulsion Explosives

Bringing into intimate contact oxidizer salt and fuel in the physical form of an emulsion and making it sufficiently sensitive to function as an explosive is the basis for production of emulsion explosives.

Ammonium Nitrate Explosives for Civil Applications: Slurries, Emulsions and Ammonium Nitrate Fuel Oils, First Edition. E.G. Mahadevan.
© 2013 Wiley-VCH Verlag GmbH & Co. KGaA. Published 2013 by Wiley-VCH Verlag GmbH & Co. KGaA.

Emulsions by definition are dispersions of two immiscible phases. The stability of such a mix depends on the thoroughness of mixing and compatibility of the dispersed phase and the continuous phase. In case of emulsion explosives, in water in oil (W/O) types, discontinuous phase is the oxidizer solution (oxidizer salts in water) and fuel blend is the continuous phase.

The two phases – aqueous and oil phase – do not mix by themselves. It is not easy to form an emulsion unless surface tension lowering emulsifying agents are used and the mixing has rapid shear action. This is well known from the cosmetics, paint, and pharmaceutical industry where emulsions are made in large quantities [3–5]. Most of the emulsions made in the cosmetic industry are oil in water (O/W) emulsions where oil is dispersed in water which forms a continuous outer phase. These are known as "greases." They are much easier to form and more stable than W/O emulsions. The water content is very high in the "grease" – almost 95–98%. However, for explosives oxidizer salts are in the aqueous phase in high concentrations and they need to come in contact with fuel intimately. This is achieved by dispersing the oxidizer phase, with the help of an appropriate emulsifying agent, as microdroplets in a continuous extended media of the fuel (oil phase).

Mere dispersion of the two phases to form an emulsion is not sufficient to make an explosive. Sensitizers in the form of micro (air) bubbles need to be present in the emulsion matrix just like in the case of water gels to impart sensitivity. Apart from this emulsion explosives may also contain ingredients such as metallic powders (aluminum (AL)), organic liquids, and substances for increasing the density and strength of explosive. It is also necessary that the emulsion formed does not separate into its individual phases in order to preserve its explosive properties. The availability of different types of emulsifiers and understanding their functions has gone a long way in the correct optimum usage of the emulsifying agent to obtain the desired sensitivity and stability.

6.3
General Composition of Emulsion Explosives

The different ingredients which make up an emulsion explosive are oxidizer salts such as AN, SN, Ca(NO$_3$)$_2$, SPC, fuels such as waxes, oils, atomized AL, emulsifiers, gassing agents, organic sensitizers, and microballoons containing air or nitrogen.

The emulsion explosives also follow the concept of oxygen balance (OB) wherein the explosive mix having zero or close to zero OB functions best in terms of performance and fume characteristics. The design follows calculation of the OB similar to water gels. A typical packaged general purpose small-diameter (GPSD) cap-sensitive emulsion could have the following composition shown in Table 6.1.

Table 6.1 Typical composition of an emulsion explosive.

Ingredients	Quantity (%)
Ammonium nitrate	62
Sodium nitrate	6
Sodium perchlorate	5
Calcium nitrate	8
Water	11
Fuel oil	3
Waxes	2
Emulsifier	1.2
Atomized aluminum	0.8
Gassing agent	1.0
Oxygen balance	Close to 0.0.

6.4
Structure and Rheology

The basis for a good emulsion explosive in all respects is the structure of its base emulsion [2, 6]. The oxidizer droplets of extremely small size are supercooled liquids. The crystallization point of the OB is well above room temperature. The salts are totally in solution when they are mixed with the fuel phase which is also heated to such an extent that solid fuels are fully melted and blended into the fuel Oil. Cooling such a supersaturated solution simultaneously with dispersion and emulsification produces droplets where crystallization does not set in even when temperature is well below crystallization point. Even when crystals are formed in due course, they are extremely fine and to a great extent separated by a barrier film of the fuel blend. Eventually due to temperature fluctuations, pressure while handling, breakage, or weakening of the emulsion crystallization does set in and crystals grow after rupturing the barrier. This leads to the loss of sensitivity of the emulsion explosive. It is therefore important to obtain a matrix of emulsion which is firm in the case of packaged products requiring long shelf life. In the case of pumpable products the emulsion should be more flowy as it has to be pumped at ambient temperature but at the same time possess a well-formed emulsion structure to have sufficient water resistance.

Rheology is the physical consistency of the emulsion during its processing as well as when it is stored as an explosive or as a matrix. Rheology (viscosity, flow characteristic) is different at different stages of manufacturing and needs to be controlled as desired in order to have a smooth processing flow. Rheology depends on composition and temperature. By composition is meant in this case the composition of the aqueous phase and that of the fuel phase. The ratio between the two influences the rheology of the emulsion. The rheology is also dependent on temperature; the higher the temperature the lower the viscosity. Emulsions are pseudo plastic in nature. They flow under pressure. All these parameters of composition and process conditions can be manipulated to obtain desired stiffness or flowability. For the same composition the smaller the solution droplet size the greater is the viscosity of the emulsion. If the fuel phase viscosity increases at a constant ratio of oxidizer to fuel, the resulting emulsion is thicker with higher viscosity. The fuel phase viscosity in turn is regulated by its composition which can range from 100 to 30% oil, rest being wax. Waxes also vary in melting point (MP) and viscosity and this also needs to be standardized.

The basic emulsion needs to be flowy so that other ingredients can be added and mixed thoroughly. Flow ability is also needed in processing till the explosive enters the package. On the other hand, it should be sufficiently viscous to hold firmly air bubbles generated in it either by mechanical mixing or by chemical gassing and not allow them to escape or coalesce while processing. In the package and in the borehole it may be desirable for the emulsion to become more viscous for purposes of transportation without leakage and offer resistance to water. Most emulsions are made without any cross-linking to from a rigid structure seen in water gels after cross-linking. Some effort has gone into the use of polymer systems (oil compatible) in the fuel phase to arrive at an end product which is rubbery and closer to the "feel" of nitroglycerine (NG)-based explosives. Such emulsions are claimed to have greater stability and sensitivity in storage and not affected negatively by temperature fluctuations.

Shear rate is another important parameter influencing rheology of emulsions. The higher the shear rate up to a certain extent the smaller is the dispersed droplet size and more viscous is the emulsion formed. Based on this behavior process parameters such as speed of mixer, the design of the mixer itself is established for the end product rheology desired which again depends on the end use conditions.

Microscopic study has revealed that the dispersed structure of the oxidizer in fuel blend resembles a honey comb structure. The thickness of the fuel blend barrier separating the oxidizer droplets is less than 0.0001 mm. This suggests that the area of contact between the oxidizer and fuel is extremely large facilitating rapid progress of the explosive process throughout the body. Further the oil/wax film encircling the droplets of oxidizer gives substantial resistance to ingress of water and leaching out of the salts.

The presence of tiny bubbles in the internal structure of the emulsion further enhances the explosive properties due to the phenomena of "hot spots" actively propagating the initiated detonation. The bubbles could be air or gas introduced mechanically or chemically by decomposition of gas generating compounds or by addition of commercially available synthetic hollow spheres of glass, or resin material. The sizes of the bubbles are in the range of 0.1 mm diameter. The effect of synthetic hollow bubbles is much more pronounced and beneficial in an emulsion explosive matrix than in a water gel. Without the need for AL flake powder to act as sensitizer, emulsions are fully sensitized by air bubbles whatever their origins as long as they are present in adequate numbers and of the desired size. The authors' attempt to obtain cap sensitivity in emulsions with the addition of 3% flake Al powders without microballoons in the matrix did not succeed as

1) there was no density drop due to the addition of the flake powder,
2) desensitization due to the oil phase dissolving the stearic acid coating on the surface of Al flake, and
3) even the air bubbles and the droplets of oxidizer in the fuel phase appeared to get distorted and coalescence of droplets was in evidence.

This resulted in the surface area of contact between oxidizer and fuel getting drastically reduced leading to loss of sensitivity to detonator initiation, but booster sensitivity was however retained. The use of AL flake powder in emulsions has taken a back seat as it brings no advantage and the addition of (coated) atomized powder of fine particulate size is sufficient for the purpose of increasing the strength of the emulsion explosive. Use of organic sensitizers in the emulsion matrix complements the addition of air/gas bubble sensitization, and very effective emulsion explosives are manufactured with monomethylaminenitrate and microballoons in the composition.

Droplets according to Wang [2] initially are 0.1 mm and above in size but in more stable emulsions they are 0.2–5 mm since the smaller bubbles coalesce. However with adequate emulsifier quality, droplets 0.01 μm in diameter exist for a length of time. The emulsion appears transparent. Droplets with the same diameter are not the only droplets formed; at any time there will be droplets of different sizes and with time they tend to grow into larger entities. The rate at which this takes place is a measure of stability (instability).

6.5
Composition and Theory of Emulsion Explosives

The fundamentals of a good explosive in terms of the intimate contact between oxidizer and fuel as well as preserving a close to zero OB is absolutely applicable to emulsion explosives, perhaps more than any other AN-based explosive. The RMs

Table 6.2 Commonly used RM in emulsion explosives.

Oxidizer salts	Fuels	Sensitizers	Gassing agents	Stabilizers	Emulsifiers	Coolants
AN, SN, CN, SPC	Diesel white mineral oil, different types of waxes	MMAN	Sodium nitrite	Lecithin	SMO	Sodium Chloride
	Vaseline	EDDN	Urea	Stearic acid	SPANS	
	Atomized aluminum	MEAN	Hydrazine Hydrate		GMO	Oxalates
	Sulfur	Nitromethane				Bicarbonates
		Air/gas bubbles	Hydrogen			Bromine salts
		Microballoons	Peroxide			

which go into the composition of an emulsion explosive are slightly different than water gels. In its fuel, it is closer to AN/FO in the sense that fuel in both is oil based.

Table 6.2 shows the RMs used in most emulsions manufactured for commercial use.

6.6
Manufacture

6.6.1
Types of Emulsion Explosive Products

1. Packaged	Large diameter – booster sensitive	General purpose
	Small diameter – cap sensitive	—
	Special applications	Permissible
2. Pumpable	Repumpable	—
	Bulk	
	Nonexplosive matrix	
3. Augured	Heavy AN/FO	—

Whatever the final product needed, a basic emulsion matrix has to be produced which can then be finished as desired. In many bulk products for on-site delivery through pump trucks a nonexplosive matrix is produced in a base plant and finished on-site.

Production of the basic emulsion consists of mixing together fuel blend and oxidizer solution at high speed. The presence of an emulsifier enables formation of a W/O emulsion suitable for creating an emulsion explosive.

6.6.2
Manufacturing Process

Worldwide the methods used by various manufacturers are more or less similar with some slight variations depending on the ease of operations and type of explosives required, the degree of sophistication sustainable continuously during manufacture by local support structure such as skilled manpower, availability of spare parts, and so on.

The methods of manufacture described basically create conditions for formation of a W/O emulsion with good stability. The water phase consisting of a solution of AN and other oxidizing salts at an elevated temperature (80–90 °C) where all the salts will be fully dissolved is brought together with the fuel phase consisting of wax, oil, and emulsifier also heated to 80–90 °C in order to facilitate free flow and easy mixing with the oxidizer. The mixed phases are formed into a premix in a jacketed vessel with low energy stirring. This step is intended to ensure homogeneity of the mix. Subsequently the premix is subjected to further shearing forces which enables a stable emulsion (nonexplosive) matrix formation. The type of final product required will decide the type of mixer and further processing steps. In general, the finishing stage consists of adding other ingredients required for imparting additional strength, sensitivity, and mixing and conveying in a screw mixer/conveyor. Before reaching the packaging equipment gassing agents are also added en route. The screw conveyor is designed to mix the material moving along to such an extent that the uniform distribution of the added material be it solid or liquid is achieved. The compositions available are several depending on the type of product required since their physical properties and sensitivity characteristic will differ.

In general, three types of manufacturing methods are practiced (all flow sheets concerned with manufacture of emulsion explosives are given in Appendix A).

6.6.2.1 Batch Process

Here all operations of blending and mixing are performed in predetermined sequence till end product is obtained in a single blender/mixer. This operation is similar to that practiced in the manufacture of NG explosives and water gels. The quantity per batch is 500–750 kg. The batch size should not be unwieldy and within the capabilities of the mixer blender as otherwise the emulsification will be problem and batch time will be very high. Since at the end of the mixing process the product is an explosive, the entire operation will have to be treated as an explosive operation and man, distance, explosive quantity limits as prescribed by statutory rules will have to be followed. Ingredients added are liquids, granular solids, and fine powders in large and very small quantities and accordingly the accuracy of the addition operation has to be adjusted.

Once the mixing operation is complete, the explosive is given up to a cartridging machine located either in the same building or in a different one nearby.

The batch process is conducive to production of cartridged explosives and is rarely practiced for bulk explosive manufacture where continuous operations are preferred. The batch process is operated where sophisticated controls for flow are not available for one reason or the other. The batch process is mostly the outcome of utilizing mixers already available and used earlier in the manufacture of AN/FO or water gels.

6.6.2.2 Semicontinuous Operation

This is a two-step process. First a matrix consisting of a high-density nonsensitized emulsion is prepared and then this is converted into a finished product which is an explosive by the addition of sensitizers and other ingredients. The matrix is made continuously and the finished explosive in a batch operation. The semicontinuous operation is also useful for the production of large volumes of bulk explosives in an efficient manner since the matrix can be produced continuously and pumped into holding tanks for transfer to bulk trucks where they are converted into explosives by adding sensitizers just before going down the borehole. Since the matrix quality can be kept high by using a right emulsifier, the matrix is fit for transport in simple nonexplosive trucks over long distances to be stored there in silos and made use of later in bulk trucks equipped for sensitizing them and loading into the boreholes.

6.6.2.3 Fully Continuous Process

As the name itself implies, here both the matrix and final products are continuously obtained. There is a substantial reduction in the quantity of explosive accumulated in the process provided that the explosive packed is continuously hauled away.

The continuous system of production can also cater to nonexplosive matrix production in large quantities at a high rate of production useful for bulk explosive delivery.

A continuous process without premix has also been developed and practiced. Here both the fuel blend and oxidizer solution are brought into contact with each other in the desired ratio in a high-speed mixer. A gassing solution can also be introduced in the mixer. The emulsion coming out of the mixer can be an explosive depending on the composition and rapidity of gassing. Provision is made to pump the matrix through an online screw mixer where solids can be added and mixed if needed and then moved to a hopper for packing.

If the flow rate of addition of RMs, emulsion transport speed, and packing speed are fully synchronized, then the operation is truly a continuous operation from start to finish. The actual holdup of explosive on line is very less in quantity. The process lends itself to sophisticated controls, automation aids, and needs very few personal to operate.

Choice of the process and the degree of automation will have to be decided taking local conditions such as availability of skilled manpower, spare parts, and so on.

The oxidizer salts are fully dissolved in water and the fuel blend is a homogeneous mix of wax and oil so that they can be pumped or allowed to flow into the mixer by gravity. Any other additive needed during the emulsification is also added and dissolved either in the oxidizer or fuel. If solids are needed to be added after the emulsion is formed, there is provision in the screw mixer for this.

Thus, it is necessary to maintain the temperature of the oxidizer, fuel blend high enough to enable free flow without choking. Even the emulsion formed is not allowed to cool on line in the continuous process. Any cooling if required is done only after the product is packed as finished cartridges. Crystallization of oxidizer salts or congealing of wax must be avoided at all costs if the emulsion is to be of good quality. Generally both OB and F/B are kept about $10\,^{\circ}$C above their crystallization and fudge points, respectively. Hence, the lower the water content in the oxidizer blend and lower the oil content in the F/B higher will be the required processing temperature. Some manufacturing processes keep part of the AN outside the oxidizer solution and to be added later in the form of solid AN (prills).

Bulk emulsion compositions contain more water both in the matrix and as finished product, and hence such a product is processed at lower temperature. Naturally in all the process lines, all the vessels have to be insulated to prevent loss of heat/energy and to avoid solidification in the lines.

6.6.2.4 Critical Equipment for Production of Emulsion Explosives

Mixing equipment and pumps are the most critical pieces of equipment used for production. Rota meters, temperature, pH, and pressure indicators are all important instrumentation devices for process control. The packaging equipment is critical in the sense that the output of the final product and acceptability by the end user is totally dependent on its optimum function though it is not directly involved in the process.

For the batch process the pieces of equipment used are similar to those used for the production of water gels. A double-helix-type mixer made of stainless steel (SS) with gland packing outside the mixer body and with provision for high/low mixing speed and reverse operations is adequate. Clearance is 3/4 in. for the gap between the blade and side walls. The mixer should be jacketed for the flow of hot/cold water.

While an emulsion can be made and was made in the earlier days by using the batch process with the above type of mixer from start to finish, the practice now is to use it only for finishing the premix and for the addition of other ingredients such as gassing agents, AL powder, AN prills, and so on.

The authors' experience of batch-produced emulsions showed that emulsion structure was not conducive to long-term stability especially under tropical conditions of storage. Also the formation of an emulsion initially was much dependent on the operator judgment. Often the emulsion would break and separate. It was possible to rescue such batches by once again heating, remixing with an additional suitable emulsifier added into the mixer. Productivity was impaired. The majority of the droplet size in the batch process was larger than usual (3–8 μm) easily visible under a simple laboratory microscope. Most producers now use a semicontinuous or continuous process for increased productivity with less holdup.

In the semicontinuous/continuous process the time for forming the emulsion matrix is extremely small unlike the batch process which can take up to 10 min for forming the emulsion. Even at a reasonable throughput of 20 kg/ min, the residence time of the premix in the mixer is half a minute or less before it gets expelled and in this short time the emulsion has to be formed fully with the desired microdroplet dispersion. This is possible only if high energy is fed into the mix and subject it to high shear forces. Homogenizers, continues recycle (CR) pin-type mixer, and jet mill-type mixers are necessary to achieve emulsification of high order continuously. It is up to the chemical engineer to select the most suitable type of mixer, but currently CR mixers and colloid mills are favored. It is the authors' suggestion, especially where full automation is not possible and control of the process is uneven, to use a premixer to make a loose homogenized emulsion blend initially and feed this continuously to the high-speed shear mixer in order to obtain an emulsion of consistent quality.

The premix operation is a low speed low energy simple mixing operation. In a cylindrical vessel OB and fuel blend at the desired ratio are fed in at the bottom. The vessel has a propeller-type agitator stretched right to the bottom. OB/FB form a loose emulsion at the bottom which gets pushed up and exits at the top to move into the continuous mixer. Feed into and out of the premix vessel is synchronized with the high-speed emulsifying mixer output. Any problem of holdup on line can be taken care of by stopping the operation of the premixer. The viscosity of the premix is less than 20 000 cps.

In a variation of the semicontinuous process the premix is made in a batch mixer with a volume sufficient to produce enough quantity of material to operate the high-speed mixer continuously till such time no more pre mix is available. If two premix holding silos are available, they can be operated in tandem and the high speed mixer can be fed continuously till all the OB and fuel blend are exhausted.

Where the bulk explosive or nonexplosive matrix is required the premix is sent not through a high-speed mixer but through a static mixer. This is sufficient to give an emulsion of required short duration life and to meet performance requirement of the bulk explosive blasting operation. The static mixer length (residence time) and the packing used can be adjusted within certain limits to obtain the desired quality of a bulk product. Finer packing and longer column length of the static mixer will give a better quality emulsion but due to pressure drop through the static mixer, output will get reduced.

6.6.2.5 Pumps

Much work has been done to arrive at the right type of pump for producing an emulsion explosive. The criterion that needs to be satisfied primarily is one of safety. The material should be safely pumped out in such a way that the emulsion should not be subjected to undue high pressure so that its inner structure remains intact. In other words, the micellar structure should not deteriorate in that smaller droplets are forced move and from larger droplets. The same criterion also applies even to gassed product. The air/gas bubble structure should not be subjected to stress/pressure to make them coalesce and become bigger in size. The pumps available satisfying the above criterion are

- diaphragm pump (single or double),
- mono pumps (positive displacement progressive cavity type),
- screw pumps, and
- peristaltic pumps.

Today the trend is to use positive displacement progressive cavity pumps with stator/rotor made of nonmetallic material and with other safety devices. These are described in detail in a separate chapter.

6.6.2.6 Packaging Equipment for Emulsion Explosives

Predominantly chub style (sausage) form fill seal machines are preferred. The features are already described in Section 5.3.3. The easy flowability when hot is a great advantage in the case of emulsion products. Packaging of emulsions in paper and rigid plastic is also done by some manufacturers using their own design of filling machines.

6.6.3
Raw Material for Emulsion

The oxidizer salts used are the same as described earlier for water gels, namely AN, SN, Ca(NO₃)₂, and SPC. The requirement on quality checks is also similar. Chemical purity, absence of anticaking additives, surfactants, foreign bodies, and contaminating chemicals like nitrites are the main criteria to be fulfilled. AN prills (low density (LD)) are also used for doping the emulsion matrix. The prill quality should be similar to that used in AN/FO.

6.6.3.1 **Fuel Blends**
Fuel component consists of oils and waxes. As regards oil there are a number of choices – synthetic origin and plant/edible oils. The availability could depend sometimes on local conditions and cost. Generally natural edible oils are avoided due to their cost and auto-oxidation. Depending on the viscosity of the emulsion finally required petroleum based oils are selected. A viscous emulsion is obtained if a thick oil-like furnace oil is used. Vaseline or white mineral oil will produce a thinner more flowy product. The reactivities of the oils and waxes used also vary. The more reactive ones are used for small-diameter emulsions to obtain better *in situ* performance. *Reactivity* is defined as the speed and case with which it decomposes and takes part in the detonation. Waxes can be paraffin/microcrystalline varieties. Higher melting waxes are used when stiffer product such as those preferred for use in SD cap-sensitive explosive is required. Microcrystalline waxes show more tendency to form and accept air entrainment and these are used in conjunction with usual regular paraffinic wax in compositions used for chemical gassing or mechanical aeration process. Microcrystalline wax has also the beneficial property of crystallizing in very small size (fine crystals), which does not reduce the sensitivity/stability of the explosive in storage.

Although both paraffin and microcrystalline waxes are from the same source, petroleum, they have widely differing properties. They are obtained from different fractions and undergo different processing. While the structure of regular paraffin wax is mostly linear with branching occurring only at the ends, microcrystalline wax has large amount of branching within its structure. The MP, molecular weight, and boiling point differ. X-ray diffraction studies have revealed that microcrystalline wax has a very fine crystalline structure on cooling as against larger size crystals of paraffin wax. Table 6.3 shows some differing properties.

Use of microcrystalline wax is highly recommended in cap-sensitive small-diameter packaged emulsions to enhance and ensure long-term storage stability and in obtaining a stiffer product on cooling to ambient temperatures. Such a product is easier to handle and load in field conditions. MCW also acts in modifying the paraffinic wax crystals to smaller size. MCW does not tend to break up the continuous phase and hence helps in forming a stable W/O emulsion provided the right type of emulsifier is used.

The fuel blend of oil and wax which make up 6–10% of the composition is selected on the basis of chemical and physical stability, rheological properties

Table 6.3 Properties of paraffinic and microcrystalline waxes.

Properties	Paraffinic	Microcrystalline
Melting point (°C)	52	64–70
Molecular weight	300–500	500–700
Boiling point (°C)	300–450	700
Lipophyllic property	Mild	Strong

during process and in final product, its storage (thermal) stability, resistance to water, and ability to participate fully in the detonation process to deliver maximum energy by combining effectively with the oxidizer. Thus fuel blend can consist of all oil, oil/wax, or all wax composition but preferred ones are oil and wax mixtures (with MCW forming a small percentage).

Although diesel oil is highly recommended as the most ideal fuel for AN/FO, it does not have the same status in emulsion explosives as it is not easily emulsified and WMO or vaseline is preferred in this respect. Even furnace oil is used to stabilize the emulsion because of its higher viscosity. However due to its higher calorific value, higher flash point, cost, and easy availability, diesel is still used in many countries in emulsion explosives. In most cases it is mixed with other oils and waxes.

Apart from paraffin and microcrystalline waxes, certain high molecular weight organic compounds produced in the petrochemical industry can be used to enhance the stability of the emulsions by addition in the fuel phase. For example,

1) Polyethylene $(CH_2-CH_2)_n$ which is a straight chain vinyl polymer with a molecular weight of 5000–20 000 can be used in the fuel phase along with paraffin wax. Higher molecular weight polymer needs to be avoided as its higher viscosity causes difficulty in emulsification. The polythene polymer is reported to have increased microcrystallinity in paraffin wax and hence facilitating lesser use of MCW. The improvement increases with the amount of polyethylene added. The MP is elevated on addition of polyethylene as well as viscosity and thus can benefit the storage stability of emulsions at higher temperature such as those prevalent in tropical countries in summer.
2) Similar benefits are claimed by the addition of polypropylene in the fuel phase.
3) Use of copolymer ethyl vinyl acetate (EVA) increases the hardness of waxes and hence can be beneficial in enhancing storage stability.
4) Other polymer additives such as polyisobutylene, copolymer of butadiene-styrene are also claimed to have beneficial effects.

Waxes from natural sources like Bees wax are reported as being used in China with good results.

6.6.4
Sensitizing in Emulsion Explosives

The need to impart sensitivity in emulsion explosives in an effective manner without employing self-explosive ingredient has given rise to different systems.

6.6.4.1 Air Entrapment or Occlusion while Emulsification by Mechanical Agitation
Every emulsion has the property to take in air/gas up to a definite quantity when they are stirred. This is especially pronounced when the emulsion is formed at higher temperatures and it is cooled with stirring. Occlusion temperature and ease of air occluded will depend on the composition of the fuel and oxidizer phase. Microcrystalline wax is known for its ability to take in and hold occluded air/gas. In the case of mechanical mixing in open mixers such as a double-helix mixer with simultaneous cooling, air is entrapped in the form of tiny bubbles whose size can range from a diameter of 0.5–100 μm. If the mixer is a closed vessel and other gases such as N_2 and CO_2 are introduced deliberately through a sparger, then the microbubbles will have the corresponding gases in their inner space. It has been reported that microbubbles having N_2 or CO_2 in them are more stable than those with air as regards growth and movement.

The temperature at which the density of an emulsion drops suddenly while stirring and cooling is going on can be identified by the change in color of the emulsion from off-white (yellow) to white, and is known as *occlusion temperature*. Crystallization of the salts on cooling also influences occlusion. Occlusion temperatures for most common emulsions for explosives are in the range of 45–42 °C. Generally in a standard emulsion with a wax to oil ratio of 1 : 1 in the fuel blend on cooling while mixing there will be a drop in density from 1.38 to 1.16 gm/cc (around 15%). A range of 10–20% (by volume) drop can be achieved depending on the composition of the fuel blend. The stability of such air in the matrix on storage is different for different compositions and will depend on the rigidity which the emulsion will attain at ambient storage and its ability to withstand thermal degradation.

6.6.4.2 Chemical Gassing
Chemical gassing has been resorted to more and more within the last decade as it lends itself to better productivity (less mixing time) in the batch process and can be used on line in continuous production plants.

The basic principle consists in adding certain chemicals to the emulsion matrix when it is being manufactured. These chemical(s) have the ability to produce gas by interacting with the oxidizer salt (AN) present in the matrix. For better dispersion the chemicals are used as solutions (aqueous) or as liquids miscible with the oil phase. The liquids are organic in nature and come from hydrazine compounds, azoamino benzene, hydrazine hydrate, and nitroso compounds. The inorganic gassing systems are preferred because of the cost and easy availability. They belong to groups of nitrites, bicarbonates, and peroxides.

Currently the gassing systems used in practice are

1) sodium nitrite/thio urea,
2) sodium nitrite/sodium thiocyanate,
3) hydrazine hydrate/sodium dichromate, and
4) sodium bicarbonate/acetic acid.

The reaction of sodium nitrite with NH_4NO_3 as shown below is utilized:

$$NH_4NO_3 + NaNO_2 \rightarrow NH_4NO_2 + NaNO_3$$

$$NH_4NO_2 \rightarrow N_2 + 2H_2O \tag{6.1}$$

However, the above reaction may proceed slowly and hence a catalyst or a promoter usually having a thiocyanate radical is used. The reaction mechanism in this case is

$$HONO + H + SCN \quad \rightarrow NOSCN + H_2O$$

$$NOSCN + RH_2N_2 \rightarrow RH_2N - NO + SCN$$

$$\rightarrow N_2 + H_2O + R \tag{6.2}$$

Reaction (6.2) is further accelerated by use of amines, and thus the system used can be

$$\frac{\text{Ethanolamine Nitrate}}{\text{Urea}} \text{ / Sodium Thiocyanate / Sodium Nitrite}$$

and to promote greater contact with AN and sodium nitrite even when AN in the matrix is protected by a oil barrier, sodium nitrite solution should contain urea. The pH needs to be acidic for rapid reaction and is adjusted to 4.5, and concentration used is up to 1% of the total explosive weight. The composition of the gassing solution is urea 2/NaCNS 2/sodium nitrite 1 or NaCNS 2/NaNO$_2$ 2. In a generally used composition for small-diameter explosives when the gassing solution is added at a processing temperature of 52 °C, the density of the matrix is brought down from 1.38 to 1.05 g/cc in 10 min time. Without use of thiocyanate the same drop in density will take more than 1 h. Thus, control of gassing rate and quantity is in the hands of the formulator and can be tailor made to suit the process and packaging parameters. Where cartridges are polythene sheathed sausages and chilled immediately after packing the gassing rate required may be different than for a batch process producing stiffer products packed in paper.

Using the NaHCO$_3$ system produces acceptable gassing rate only at higher temperatures (65 °C and more).

6.6.4.3 Hollow Particles

In order to avoid the uncertainties in generation of the desired size and number of bubbles at the right time, use of hollow particles containing air or gas has been resorted to. The usage of hollow particles for sensitizing the emulsion matrix has increased as the initial problems of pressure desensitization seem to have been overcome by using very thin walls in the hollow spheres. By producing larger volumes of microballoons which are also finding use in other applications the cost is being brought down to affordable levels for use by the explosive industry.

The different types of hollow particles offered on the market are

1) glass microballoons
2) resin-based product
3) perlite.

The glass product first produced in the USA by 3M Company is a standard product of bulk density 0.1–0.4 g/cm^3 and a uniform size distribution between 10 and 160 μm with majority being in the 60–80 μm range.

About 1–3% is the amount used depending on the sensitivity level required. Another product from USA Ecospheres is slightly larger of size 50–200 μm.

The resin-based products, Expancel, originates from phenol or urea formaldehyde resin base made in the form of hollow spheres and containing air/gas. A product of this type has been offered with a size of 30 μm and a bulk density of 0.03 g/cm^2.

The perlite product recommended for use in explosives is an expanded particulate material. It is a product based on SiO_2 and the particle size of a fine powder on offer is around 100–150 μm. Synthetic perlite spheres contain isobutane gas and shell is a fraction of micron thick.

The beneficial effect of using expanded varieties of perlite in various mesh sizes and concentration has been claimed in China. Some of the bulk explosives use perlite with some benefit of resistance to hydrostatic pressure by keeping the air bubbles protected. Whereas failure of detonation occurred at 2.08 bars of hydrostatic pressure in the emulsion explosive without perlite, an emulsion explosive containing 2% perlite was able to detonate even at 4.3–6.0 bars pressure.

Perlite is a generic term for naturally occurring siliceous volcanic rock. When heated it expands 4–20 times its original volume due to the presence of 2–6% water inherent in it. Tiny bubbles are formed during expansion.

Vermiculite from expanded MICA is another low bulk density product which has been tried out. However, due to its open structure it did not impart cap sensitivity even at 5% level and provided only stabilization of density achieved already through aeration or gassing.

Currently, in order to obtain acceptable cost/benefit, use of a gassing agent and a small percentage of hollow microballoons is resorted to. While the gassing provides a substantial drop in density and count of air bubbles is high, use of microballoons even in smaller percentages ensures long-term stability and high velocity of detonation (VOD).

6.6.5
Crystal Habit Modifiers

It is found beneficial to influence the crystal structure of AN as it comes out of an oxidizer solution when emulsion cools. Under normal circumstances it is likely that AN will tend to form larger crystals, but when other salts such as $NaNO_3$, sodium per chlorate, and $Ca(NO_3)_2$ are present, the crystals coming out are smaller and somewhat rounded in structure. The use of a crystal habit modifier for AN is a well-known and researched subject. Use of acid magenta in dynamite has been known from early times. The usefulness of crystal habit modifiers is more important when the AN goes in and out of a solution due to fluctuations in storage temperature. The AN getting recrystallized should not break down the emulsion structure by disturbing the droplets and media barrier by growth and shape. This can be prevented by using small quantities of a crystal habit modifier. These are generally water soluble surface active agents and added to the oxidizer solution prior to emulsification in percentages of up to 0.5% of explosive weight. As these are highly active surface active agents, care should be taken about the quantity added. Larger quantities surely interfere with emulsion forming. Compatibility with chemical gassing systems being used needs to be checked. Amongst the various types of surface active agents available (cationic, anionic, and amphoteric) anionic surface active agents give the best results. The $C_{12}-C_{18}$ alcohol-based alkali metal sulfonates, phosphates, phenyl or napthyl alkali metal sulfonates, and sulfates can be used effectively. The use of crystal habit modifiers is particularly useful in low water compositions.

Other compounds reported as useful crystal habit modifiers are sodium lauryl sulfate (SLS)/dodecyl sodium sulfate/sulfonate (DSS).

6.6.6
Emulsion Promoters

Highly chlorinated paraffinic hydrocarbon has been discovered to make emulsification easier thereby producing a finer dispersion – smaller droplets of the dispersed phase are enabled, which leads to better long-term storage stability of the emulsion and also enhances performance of explosives (sensitivity and propagation). The chlorinated paraffin is added up to 1% along with the regularly used emulsifier. Long chain $C_{10}-C_{20}$ paraffinic hydrocarbons with greater than 50% by weight of chlorine have been claimed to be more beneficial. Greater the degree of chlorination, stronger is the promoting action. With the use of high energy homogenizers and emulsification machines the benefits of using an emulsification promoter become less important once the primary emulsifier is chosen carefully to perform an effective emulsification.

I would think that for rescuing separated batches (reworking) use of the emulsion promoter would give the maximum benefit.

6.6.7
Emulsion Stabilizers

Once an emulsion has been formed, for storage stability it needs to be kept in a condition which does not show reduction in terms of its explosive properties. Certain substances such as kaolin, talc, stearates/stearic acid/sulfur (size should be less than 1.0 μm (0.005−0.5)) when added as fine powders have been found to prevent separation of phases in an emulsion and hence have been used in the pharmaceutical industry. Even fine metallic powders are found to have a stabilizing effect provided that they do not react with other components especially water present in the dispersed phase.

Phosphatide compounds have also been reported as effective stabilizers. Soya lecithin [7] is available easily and at reasonable cost and has been claimed to be effective even at 0.5% when added along with the regular emulsifier in the fuel blend. Lecithin is insoluble in water and hence can be emulsified. Lecithin can exist as α or β varieties. The latter is also termed *Cephalins*. The ratio of α to β varieties in the commercial lecithin may influence significantly its usefulness as stabilizer in an emulsion. Not enough data are available in this regard for emulsions made for explosives application. Generally lecithin commercially available has 60% phosphatides and 40% fats/oils.

The beeswax/borax combination has also been reported as an auxiliary emulsifier/stabilizer. Borax needs to be used to counter the acidity in beeswax. 0.3% beeswax is recommended along with 0.4% of borax. Excess of borax should be avoided as it may start reaction with AN. Use of beeswax in commercial products has been very limited as availability except in China is low.

6.6.8
Emulsion Chemistry and Understanding Emulsifiers: Key to Good Emulsions

Of all the RMs used for making emulsions, the most critical one is emulsifier(s). Without the emulsifier functioning the way it is needed, there will be no possibility of forming an explosive since a mere mixture of aqueous and oil phases will not be enough to make an explosive even though they may be in the right proportion. For an emulsion to function as a good base for an explosive, there needs to be an extension of the surface area of the oxidizer and fuel so that they are in intimate contact with each other. This is possible only if an emulsion is formed in which there is a finely dispersed phase and a continuous phase. Oxygen balance requirements have predetermined that in an explosive there will be a limitation of the ratio between oxidizer solution and fuel blend which is 15 : 1 approximately. Thus, the aqueous phase is available in large quantity and one would expect that an oil in water emulsion where the aqueous phase forms the continuous phase would be easy to achieve and would serve better the purpose of close contact between oxidizer and fuel. While there exist such emulsions called "*Greases*," these lack the water resistance required in an explosive especially bulk loaded ones where the explosive could come directly in contact with moisture or water in the borehole. The only way out therefore was to devise water in oil emulsions so that the outer continuous layer of oil/wax would protect leaching of inorganic salts by water. Further complications also arise due to the presence of a large percentage of oxidizing salts in the water phase because of the formulation requirements. It is only at around 65−80 °C depending on the salt/water content all the oxidizer salts would be in solution and the aqueous phase will then be a homogeneous entity

Table 6.4 HLB values.

Emulsifier	HLB	Emulsifier	HLB
Oleic acid	1.0	Sorbitan monooleate	4.3
Sorbitan trioleate	1.8	Propylene glycol monolaureate	4.5
Sorbitan tristearate	2.1	Sorbitan monostearate	4.7
Propylene glycol monostearate	3.4	Diethyleneglycol monostearate	4.7
Sorbitan sesquioleate	3.7	Sorbitan monopalmitate	6.7
Glycerol monostearate	3.8		

ready to emulsify with the fuel phase which also is made into a homogeneous melt by raising its temperature well beyond the fudge point for that particular composition. The oxidizer salt coming out of a supersaturated cooled solution and the fuel phase solidifying should not destabilize or break the emulsion in order to preserve the integrity of the explosive. This is possible only if the emulsion is formed by use of an appropriate emulsifier capable of forming an emulsion suitable for use as an explosive by satisfying the criteria of storage stability and performance.

Thus, having established the paramount importance of the emulsifier we need to look into all the aspects of emulsion physics and chemistry in order to arrive at some basic methodology to select the most appropriate emulsifier for a given composition.

Emulsifier chemistry and emulsions are well-known subjects and extensively researched and published. Classical books exist on this subject authored by Becher [8] and Sherman [9]. Griffin [3–5] has adopted his knowledge in emulsions to formulate stable emulsion explosives. Although it is beyond the scope of this book to go into details of emulsion chemistry, it is necessary for the explosive formulator to understand basic principles of functioning of emulsifiers in forming a W/O emulsion to enable him/her to use the optimum emulsifier. The pioneering work already done in this direction has narrowed down the options greatly and more or less the usage both in quantity and specific type has become standardized. It is however in O/W emulsions the choice and options are much more and scope exists in making tailor-made mixes or use new emulsifiers put on the market.

Basically the choice of an emulsifier for W/O emulsions is limited by the hydrophilic–lipophilic balance (HLB) values of available products (compounds). It is known that lower HLB values favor the formation of W/O emulsions. Table 6.4 gives values of HLB for some commonly used emulsifiers.

It is noticed that most of them are nonionic surfactants.

It is possible to blend them together if need be to cater to specific requirements brought about by the composition of the oils and waxes being used. The HLB values can undergo slight changes under the influence of certain additives used in an explosive as well as temperature and this needs to be kept in mind. Being nonionic pH and hardness of water will not affect the functioning of these surfactants.

Coordination complexes are found to enhance the capabilities of emulsifiers, and it has been reported that the addition of long chain compounds with surface activity and capability of complex formation with emulsifiers is beneficial. These can be fatty alcohols (C_8 and above) and fatty amines.

Mixed emulsifiers take care of process conditions and RM variations. For example, sorbitan monoleate (SMO) with dodecyl sodium sulfate can be used (SMO in oil phase and DSS in aqueous phase). The combination is claimed to ease emulsification and also impart stability to droplets of dispersed phase.

Over the years the explosive industry has settled to make SMO the core emulsifier with some others being used in conjunction in small quantities. Unsaturated substituted oxazolines and derivatives thereof are claimed to be very suitable as an additional emulsifier and thermal stability enhancer in emulsion explosive compositions containing high percentage of calcium nitrate. Commercial Alkaterge-T containing 60% by weight of oxazoline is useful as an active surface agent for mineral oil. Even at 0.1% it reduces the interfacial surface tension of water by 95% to less than 2 dyn/cm. It is nonvolatile and very slightly soluble in water but soluble in most organic solvents and mineral oil. It can also function as a primary emulsifier in a W/O emulsion. Benzyl alkyl dimethyl ammonium chloride and coco diethanol amine are also additives to be used together with SMO for improved sensitivity of small-diameter explosives.

It is important to know that all the emulsifiers used for making W/O emulsions substantially deteriorate in their activity when stored at elevated temperatures. In practice in fuel blends at 80 °C, studies revealed that in 24 h nearly 30% activity as measured in emulsification power had been lost due to chemical decomposition and transformation due to change in the HLB value. Thus, the best results are obtained if emulsifiers, stabilizers, and promoters are added to fuel blend just before emulsification or at least their residence time is kept less than 4 h at high temperature (80 °C).

Span 80 (SMO) is obtained by reacting sorbitan with oleic acid using alkali (NaOH) as a catalyst at elevated temperature. Dehydration is a must and controlled conditions are necessary as esterification produces not only monoesters but also di- and tri-esters. It is the mono oleate which is most useful as an emulsifying agent and hence production of maximum amount of monoester is required. Inert gases such as N_2 are used during the reaction. Parameters for control are rate of heating, inert atmosphere, and dehydration.

Check on quality of SMO is done through measuring acid no., saponification value, iodine value, and hydroxyl number. There are standard methods being used by a laboratory chemist for experimentally obtaining these values.

Acid No.: Alkali (KOH) necessary to neutralize free fatty acid (oleic acid) in the emulsifier

$$\text{Acid No.} = \frac{N \times V \times 56.1}{G}$$

N = concentration of the KOH standard solution

Table 6.5 Specification for SMO.

Acid number	Less than 7
Iodine value	60–75
Saponification value	150–160
Specific gravity	1.00
Hydroxyl number	200–220
Freezing point	Less than $-10\,^{\circ}$C

V = ml of KOH required (titer value)

G = sample weight.

Saponification value.: Alkali (KOH) necessary to neutralize free fatty acid and saponify ester (combined acid) group in 1 g sample of emulsifier determined by titration using hyposulfite and starch as indicator.

Hydroxyl No.: Alkali necessary to estimate acetic acid remaining after acetalyzing the emulsifier.

Table 6.5 gives these values for commonly used SMO.

Polyisobutylenesuccinic anhydride (PIBSA) and mixed emulsifiers are also claimed to form excellent stable emulsions of W/O type. Mixed emulsifiers consist of a conventional emulsifier like SMO together with a synthetic polymeric emulsifier which is amphipathic in nature. Amphipathy is defined as the property of a polymer with molecular weight 500–1000 having one segment soluble only in oil phase and another only in aqueous phase. These are polyalkylene glycols. Preferred is polyethylene glycol. Together with the mixed emulsifier some amount of phosphatide (0.5–1.0%) is also very beneficial. The ratio of the mixed emulsifier to phosphatide is 1.3 : 1.5.

Stearic acid and stearates are also good emulsifiers. Although their HLB is way over [10], it is still able to form a W/O matrix which gives an emulsion explosive of very good stability. Quantity of stearate like Na stearate can be used at 1–4%.

In spite of the various possibilities outlined above current commercial emulsion explosives used for bulk loading (pumpable) having low shelf life employ only SMO that too not necessarily of the highest purity (mono ester content). It is only in packaged product where longer storage life is required combinations and special mixtures are used.

The action of an emulsifier is based on it being a surfactant. The function of a surfactant is to lower the surface tension of the liquid to which it is added. Surface tension itself can be defined as the force acting upon a liquid at its surface resulting out of the difference in force acting on a molecule at the surface as compared to the molecule in the body of the media.

In case of emulsions formed between two liquids interfacial tension (IT) becomes important. The emulsion explosive consists of two phases – the aqueous and the oil phase. The aqueous phase is not pure water. It contains a considerable amount of solids dissolved in it which increases the surface tension of water. Generally fatty acids of high molecular weight reduce the surface tension of H_2O. The presence of polar groups is responsible to reduce the surface tension.

The surfactants are classified as (i) nonionic, (ii) cationic, (iii) anionic, and (iv) ampholytic.

Nonionic surfactants are those most effective in the emulsion for explosives and hence mostly used. Anionic surfactants such as sodium alkylbenzene sulfonate are also finding some use in explosives (as crystal habit modifier and additional emulsifier). Using molecular weight is another method to classify surfactants. Lower molecular weight surfactants with molecular weight ranging from 200 to 1000 are SMO, glycerol monostearate, lauryl alcohol, and sodium sulfate and are used in explosives.

The hydrophobic and hydrophilic positions of a nonionic surfactant can be modified to suit its role as an emulsifier in a particular situation. The nonionic surfactants are classified based on their structure:

1) *Polyglycols polyoxyethylene types* – formed by reaction of alcohol or phenol with ethylene oxide (Tween series). The number of ethylene oxide groups added can be varied. Higher the ethylene oxide groups present, higher is the solubility in water.
2) *Polyatomic alcohol types* – based on alcohols such as glycerol, sorbitol, and so on.

6.6.9
Concept of HLB and Its Use in Emulsification

Use of HLB for evaluation of nonionic surfactant (Griffin, Atlas) [5] is according to the equation

$$HLB = 7 + 11.7 \log \frac{M_W}{M_O}$$

M_W = molecular weight of hydrophilic group
M_O = molecular weight of hydrophobic group.

Stronger polar group gives higher HLB. Water solubility is a simple test. No dispersion in water indicates a low HLB value 1–4 and is more suitable for a W/O emulsion, poor dispersion is seen for emulsifiers of HLB 3–6, and clear solution in water is seen for compounds with HLB > 13.

There are many methods for determining HLB values. ATLAS method is applicable to nonionic surfactants such as polyoxyethylene derivatives and polyhydric alcohol fatty acid esters. Griffin of Atlas [3–5] uses the formula

$$HLB = 20 \left(1 - S/A\right)$$

where
S = saponification number
A = acid number.

The upper and lower limits of HLB are set at 20 and 1. Thus for glycerol monostearate $S = 161$, $A = 198$, and hence HLB is 3.8.

HLB $= (E + P)/5$ is used for those fatty acid esters of Rosin oil, lanolin

$E = $ wt% of oxyethylene content
$P = $ wt% of polyhydric alcohol content.
where ethylene oxide is used, HLB $= \frac{E}{5}$.

The above is the simplest way of determining HLB.

The method has certain limitations but works well for nonionic surfactants which is of interest for emulsion explosives. Griffin also expounded the theory that emulsifier efficiency reaches a maximum (reciprocal of consumption) as given by the peak attained in a bell-shaped curve. The range of operation of an emulsifier is given by the area of the bell-shaped curve. Broader the area, greater is the variety of phase systems and conditions of pH, and temperature covered by the emulsifier. This can be achieved by mixtures of similar types of surface active agents.

The HLB value is affected by the molecular structure of the emulsifier. This could be molecular weight, chain length, position of groups, and effect of branching. An important fact arising out of this is that apart from the HLB value, the affinity between the substances being emulsified and the hydrophobic group of surfactant is crucial. If the affinity is weak, surfactant will separate from the emulsified particle and the phase will separate. Micelle formation will result. For strong affinity the molecular structure should be similar. Thus, surfactants based on fatty hydrocarbon molecules are better when mineral oil is used in the fuel phase.

Recapitulating, an emulsion is defined as a system of two immiscible liquids wherein one is dispersed in discrete microdroplets in the other. The droplets are referred to as the disperse or discontinuous phase and the outer phase as a continuous phase. To keep the dispersed phase in a stable condition and even to facilitate the formation emulsifiers are needed. Emulsifiers reduce the surface tension of the interfacial contact area and stabilize the emulsion. Emulsions are classified as O/W or W/O systems. Natural emulsions such as milk is O/W type. Natural milk has droplets of different sizes, is unstable, and separates out eventually. To stabilize milk it is homogenized into uniform particles. Butter is a typical W/O emulsion. Most cosmetic and medicinal creams are O/W emulsions. Synthetic emulsions are manmade O/W types. Emulsions made into explosives are W/O type. Physical and chemical properties of an emulsion are different from those of the individual components.

The easiest way to identify the type of emulsion is to check the solubility of the continuous phase in H_2O. A marker dye can be used for easier visual check.

Dilution or miscibility of the continuous phase in water is another check. If the emulsion blends with water easily, it is an O/W emulsion. Conductivity measurements are also useful in determining the type of emulsion. W/O emulsions with oil in the outer phase are poor conductors of electricity. The O/W systems especially if the water has electrolytes are good conductors. This fact is also utilized as a quality check tool. The W/O emulsion tending to separate will show more conductivity than its original state.

Fluorescence under UV is another way of checking the type of emulsion. If it is W/O the entire body of the emulsion shows fluorescence.

Simple dipping of filter paper in an emulsion gives a quick check. If it comes out wet, the emulsion is O/W.

6.6.9.1 Effect of Factors on Stability of Emulsions

The HLB value has a major influence on the type of emulsion formed. As mentioned earlier, HLB 3–6 favors W/O type. However, on many occasions one cannot have such sharp limits of HLB. For example, fatty acid polyoxyethylene ester of HLB = 8 can give an O/W emulsion at ambient temperature but inverts into W/O at higher temperatures ($>35\,^{\circ}$C). At still higher temperatures separation of the two phases takes place.

Steric hindrance and position occupied by the hydro and lipophilic groups affects the type of emulsion formed. Wang [2] quotes the example of Na and Ca oleates where Na oleate gives O/W type but calcium oleate gives W/O type.

The difference in interfacial tension between an emulsifier and each of the individual phases also influences formation of the emulsion. $A < B$ favors formation of W/O, where $A =$ IT between oil and emulsifier film, $B =$ IT between water and emulsifier.

Mechanical agitation done in a particular way, sparging, can lead to inversion with certain emulsifiers. A container wall material also plays a part. Glass walls are more suitable for O/W and plastic for W/O emulsions.

Physical Chemical Influences Temperature reduces the viscosity of the phases and facilitates emulsification, but too high temperature (specific to each system) can cause instability. Further HLB values undergo a change with temperature. Their effectiveness decreases with the rise of temperature. HLB values in a lower range 3–6 change upward with the increase in temperature. In formation of emulsion explosives high temperature is a must in order to keep the oxidizing salts in solution; it is only on cooling with mixing of the two phases that emulsion formation takes place.

Droplet size and distribution determine the quality of the emulsion needed to make a stable and sensitive explosive. The smaller the droplets of the dispersed phase, the more stable the emulsion. The size can be between 0.2 and 0.5 μm. Smaller droplets in pure emulsions tend to grow faster due to increased movement (Brownian movement), but if stabilizers are added they can be made to grow slower. The droplets are not of the same size. A narrow range is helpful. The emulsions with a narrow high peak of bubble size and number show instability when their distribution moves toward a diffuse broad range. Droplet size and distribution are controlled by the emulsifier, its quantity used, and type of agitation. The Schwarz and Bezemer [11] formula for calculating particle size and distribution in emulsions has been developed from experimental data, but microscopic measurements can be resorted to for practical purposes. Physical appearance itself is another indication.

At one end is macro globules seen by naked eyes, indicating a bare minimum of emulsification. At the other end is a transparent product with droplets less than 0.5 μm. The viscosity of an emulsion is important during manufacture. Emulsions need to be moved from the mixing vessel to the packer in a packaged product; in a bulk product, they need to be stored in mass, moved to the delivery truck, and from the truck pumped into the borehole. One would like to conduct all these movements once the emulsion has cooled to ambient temperature without further reheating or undue pressure. The behavior of viscous liquids and emulsions when subjected to shear stress (pressure) has been extensively studied in chemical engineering. Most emulsions exhibit what is known as *non-Newtonian character* where the viscosity value is a function of the shear rate (more the pressure or shear applied, lower is the viscosity at that point).

The apparent viscosity is, to a great extent, dependent on the viscosity of the external phase. Complex mathematical equations are available for bringing out the relationship between the viscosity of the emulsion and that of its phases, volume (concentration) ratio, IT, droplet size and number, and electrical charge [8, 9].

Cooling of Emulsions Much effort has gone into the investigation of the effect of cooling after the emulsion has been made and packed. Since speedy packing in a Chub packer requires easy flowability, the emulsion mass is kept at a temperature above the fudge point (60 °C) before packing. The cartridges filled with hot emulsions if put into packing cases and sealed take a long time, even beyond 48 h in hot climate before they can attain ambient temperature. During this time it is possible for the micro air bubbles to grow into the larger size, and this can affect the performance of the emulsion explosive. Further studies revealed that such a slowly cooled product had a mushy consistency and tended to show a crust formation. The crust was continuous and on the entire surface of the explosives in the cartridge and consisted of AN crystals. The appearance of the crust was noticed in one to two months time after production/packing and thereafter the emulsion progressively deteriorated in its stability and tended toward separation with corresponding unacceptability in performance tests such as VOD and continuity of detonation (COD).

Even if the emulsion packed cartridges were allowed to cool outside the packing cases in ambient atmospheric temperatures, the phenomenon of encrustation was noticed but to a slightly lower extent. The shelf life of the explosive was also affected somewhat.

It was found through experimentation that rapid chilling of the hot cartridges coming out of an online packing machine by plunging them into ice cold water produced best results as regards product texture, stability, and performance. Crust formation was minimal even after six months of storage and shelf life with acceptable firing characteristics was extended even up to one year.

No doubt the residence time in cold water giving the best results commensurates with logistics and product movement out of the building has to be worked out as it will be composition specific. Product shelf life required is also a determinant for the extent of cooling necessary to ensure its adequacy.

The authors' experience showed that up to two months of storage the VOD of emulsion explosives with and without chilled cooling was differing only slightly (less than 300 m/s). It can be opined that if shelf life requirement is three months, then chilled cooling does not bring such an advantage in the product considering the investment cost for chilling rapidly the cartridges coming out at the rate of 100–150 numbers per minute. However, if longer than three months shelf life is required, a chilling procedure would certainly be beneficial to product quality.

Optical Properties Physically W/O emulsions are translucent milky mobile liquids. They tend to become transparent if droplet size becomes extremely small. One can use the refractive indices of disperse and continuous phases for the correlation between physical appearance and droplet size. Interfacial area can also be measured using light transmission through the emulsion and measuring the absorption.

Electrical Properties When oil is the continuous phase, the conductivity is low. This increases with more water droplets separating out from the discontinuous phase. Hence, conductivity can be used as a measure of the status of the emulsion.

The measurements are easy if the water contains dissolved salts. If an emulsifier or an emulsion is placed in an electrical field, charged droplets move to the electrode of opposite polarity (electrophoresis). Measure of the dielectric property is possible and this can be used as a quality check in the plant and later in storage for the condition of the emulsion and the probable shelf life left over can be predicted reasonably well.

Emulsifier in Composition It can be present in the oil (fuel) phase or in the water phase. Obviously for (W/O) emulsions, oil soluble/miscible emulsifiers are best added in the fuel blend. The HLB determines emulsifiers for most W/O emulsions. These emulsifiers are water insoluble and can give stable emulsions only when added as a component of the fuel blend. No doubt stability is also influenced by the mixing process.

Decomposition and loss of specific activity in an emulsifying system occur if the emulsifier is heated for long periods. If kept for several hours or days at 80–90 °C, the active ingredients in the emulsifiers will deteriorate and will not function efficiently. It will also lead to unstable emulsions. It is recommended that the emulsifier be added in the F/B just before the F/B is used for mixing with OB. Practical limitations need to be considered and the exposure to heat kept to the minimum.

Oxidizer Salts in Emulsion The role of oxidizer salts in emulsions is more or less the same as in water gels. It is to provide oxygen internally for the fuel to react. AN forms the major constituent because of its cost, availability, and thermo chemical properties. However, other salts are also needed in order to enhance the oxygen availability, increase the density of the oxidizer solution, (and of the emulsion) bring down the crystallization point of the oxidizer blend to lower the processing temperatures and save energy. Na NO_3, sodium perchlorate, and $Ca(NO_3)_2$ all come in this category of additional salts in the OB. MMAN is another compound which brings in all the desired properties in the emulsion and in addition functions as an additional sensitizer. Calcium nitrate also helps emulsification and stabilization at higher temperatures. It also contributes to cold temperature sensitivity of an emulsion explosive. These salts when added to oxidizer blend can also act as crystal habit modifiers.

Fuels other than wax/oil/emulsifiers can be sulfur and aluminum, both of which also help in energy release due to their reactions with oxygen-forming solid residues. Sulfur in combination with $NaNO_3$ can be quite effective. Both sulfur and aluminum need to be added as fine particles. Sulfur is also reported to stabilize the emulsions.

Nitric acid itself is an excellent source of oxygen. In the early 1960s, it was used by Atlas Chemicals, USA, to make some very powerful slurry explosives, using a synthetic polymer of vinyl ether/maleic anhydride as a thickener to prevent separation. However, use of concentrated HNO_3 brought its own problems such as corrosion of equipment, safety hazards in storage[1] especially if HNO_3 separates out as also the health problems associated with inhalation of nitric acid fumes.

Further, these HNO_3-based explosives could react with sulfur present in sulfide ores as they were suspected in incidents of reaction with Cu sulfide ore in copper mines leading to premature explosion. Hence HNO_3-based explosive products have been phased out of commercial use.

6.6.10
Polymer Systems in Emulsion Explosives

Emulsion explosives where number of polymers have been used to produce rigidity in the emulsion matrix in order to lock in the air bubbles and to counter act the changes brought about by fluctuating temperature are reported in patents (11 = 16). They claim that in order to prevent the mobility of air bubbles which may result in their growth or escape resulting in the reduction of the explosive properties of the emulsion, it is necessary to control the rheology of the emulsion matrix right from the stage of mixing of the two phases (oxidizer and fuel) and subsequent emulsification. It is also claimed that a suitable polymerizable material be added along with the fuel phase and this compound is polymerized *in situ*. Further variation of this concept is the addition of a cross-linker to act on the polymerized fuel phase and bring about the desired rheology both while processing and in package. These measures will also ensure a final emulsion structure which is rigid and locks in the air bubbles. Thus, long-term stability of the emulsion as regards density and number of hot spots in the matrix is ensured. Hydroxy terminated

---

1) Explosion in 1972 at Rourkela, Orissa, India (suspected due to leakage of HNO_3 from the gelled slurry, reaction with dry packing material causing fire and eventually explosion in a magazine).

polybutadiene and butadiene–styrene systems are used and polymerized *in situ* using multifunctional isocyanates.

Another recommended system to be used is maleic adducted polybutadiene which is then cross-linked using compounds of multifunctional, hydroxy group such as ethylene glycol, triethanolamine, and so on. These systems appear to function very well in compositions containing large amount of $Ca(NO_3)_2$.

A different combination is a polymerizable polyester and a polyol catalyzed by benzoyl peroxide.

Yet another system is oil soluble polyacryl methacrylates used in the fuel phase to control the viscosity.

Using alkyd resins, affecting polymerization by irradiating with UV light is claimed to give emulsions very stable to freezing and thawing cycles.

Elastomers such as latex, polybutadiene, and polymethyl acrylates are also claimed to bring in stability to emulsions.

Ethylene homo polymers or ethylene vinyl acetate copolymers are also mentioned in another patent.

The commercial usage of the above systems is in limited quantity and restricted to special applications.

6.7
Quality Checks

6.7.1
Raw Materials

RMs are checked before making up the blends. Oxidizer salts are checked for purity, moisture, absence of nitrite, and foreign materials. In emulsion manufacture like water gels/slurries, AN melt is conveniently used. The AN to water content ratio is 80/20 or 93/7. The heat in the melt is utilized for making the blend. If one has to start with solid AN, the negative heat of dissolution of AN will require substantial extra heat to be supplied for dissolving the AN and raise the temperature of the oxidizer blend.

- *Waxes*: Prechecked for MP and molecular weight, well-known sources are used.
- *Oils*: Known sources test certificate is accepted. Viscosity, MP, or freezing point is checked.
- *Emulsifiers*: Saponification value, hydroxyl number, iodine number, and acid value are checked. The HLB value is calculated and checked against the product specification. For example, if it is glycerol monostearate, HLB should be close to 3.8.

If necessary, an emulsification test can be performed using plant OB and process conditions but with a laboratory mixer.

- *Microballoons*: Check sieve size and particle diameter, sphericity. Usually since limited suppliers are available for this RM and they are of good repute, their test certificate can be used. Bulk density check can be done occasionally. Tap density gives a higher value.

- *Gassing agents*: $NaNO_2$, urea, peroxide, and hydrazine hydrate are all subject to purity and other specific tests for these are usually performed in a chemical laboratory.

6.7.2
Process Audit

This is very important and should not be neglected as it has a direct bearing on the final product quality.

Oxidizer blend: Blend is made using weighed quantities of RM as per composition. The same composition is also made first in the laboratory with accurately weighed ingredients and its crystallization point is determined. Here the OB is taken in a small beaker, say 150 ml capacity, and it is cooled with constant slow stirring. At some temperature the solution becomes cloudy due to crystallization. This temperature is known as *crystallization point* and it is unique for that particular composition of the oxidizer blend, for example a blend giving 62.5% AN, 8.0% SN, 15% Water in the explosive will have 59 °C crystallization point. Less water and more AN increases the crystallization point.

AN	62.5
SN	8.0
H_2O	15.0

In the explosive it has a crystallization point of 59 °C. Less H_2O and more AN increase the crystallization. It is quite usual that during production OB remains at elevated temperatures for many hours, even days, and there will be loss of H_2O which can be seen in higher than specified crystallization point. This has to be adjusted by adding makeup water. If while making OB, excess H_2O or lower AN is present, then adjustment has to be made after determining the AN and water content.

In a similar fashion the F/B is checked for its fudge point which is also unique for that specific composition of oil and waxes. The fudge point is the temperature at which first signs of solidification occur as the F/B cools. This is determined before the addition of an emulsifier and other ingredients in the fuel phase. The viscosity of the fuel blend can also be checked at a specified temperature but usually not necessary if composition is correctly made and fudge point is within specification. Usually the ball drop method is used in the plant for determining the viscosity of the fuel blend.

Apart from crystallization point, the density of the OB is another reference number for checking the composition. Again a carefully made OB composition in the laboratory is used to give the correct OB density at a particular temperature which is then used as reference to check the OB made in the plant. Calibration of flow meters must be routinely done (for continuous process plant) since it is very important that OB and F/B are fed into the mixer at exactly the desired ratio as per composition. In case a premix is made initially the flow of the premix to the homogenizer or colloid mill needs to be determined and regulated so that an emulsion matrix of the desired quality is formed.

Temperature controller needs to be checked for its accuracy. Both OB and FB need to be kept close to the approved process temperature.

The viscosity of the premix as determined by the ball drop method or by timing the flow through an orifice (funnel) is a good check in the plant for ensuring that the desired quality premix is moved into the emulsifying machine.

The emulsion matrix is checked for temperature and density against specification. In case of the batch/semicontinuous process, the matrix is cooled before being fed into a mixer where a gassing agent is added. The fall in density is monitored and time of mixing fixed for regular production. In a continuous process, a flow meter is used for maintaining the desired rate of addition of a gassing agent solution. The rate of flow of gassed emulsion is controlled by a roto pump to match with the packing rate when a chub machine is used. In case of other types of packers the check is only for density and weight of finished cartridges. Occasionally diameter is also checked in chub products. The density is measured as cup density usually but it can also be obtained by weighing the material in a cartridge and dividing it by the volume of the cartridge.

Check on pH of OB is made on a continuous basis in order to ensure correct pH during emulsification and afterward before the addition of a gassing agent. Solid addition of microballoons, prilled AN, sulfur, or atomized AL is done in preweighed quantities for the batch process and for the continuous addition a screw feeder is used. The rate of addition is monitored through screw velocity (screw pump speed). The rate of addition has to match the requirement of quantity as per composition.

Clipping and sealing of the cartridges are checked manually frequently on a random basis. End closures in a paper or rigid polythene cartridge are checked visually.

VOD, COD, Gap Test, and strength measurements are performed. Test methods are the same as in water gels and described in Section 5.4.5–5.4.8.

6.7.3
Special Tests for Emulsions

The tests which are more used for evaluating the quality and condition of emulsion explosives are as follows:

1) Freezing and thawing test for measuring the stability of emulsion to temperature fluctuations.

   For this test the emulsion cartridges are stored at $0\,^{\circ}$C for 4 h, then brought up to $+40\,^{\circ}$C and stored for 4 h (represents one cycle), and finally the emulsion is visually checked for separation and crystallization. The checks are done after bringing the emulsion to room temperature (25 °C). More cycles are performed and the emulsion explosive performance in VOD tests is measured up to the stage when the explosive shows failure. Crystallization is also checked by finger pressure on the emulsion. Generally emulsions withstanding eight cycles as performed above are considered good for long-term stability.

2) Water resistance is also very important as the emulsion explosive is used in large volumes in bulk loading and come directly in contact with water in the borehole. The tests are performed as described for water gels/slurries described earlier in this book.

3) Dielectric measurements are made on emulsion product/matrix to evaluate stability. Stored samples of emulsion show higher conductance if the W/O emulsion tends to break up. The dispersed phase which is an aqueous salt solution is a good conductor as compared to the fuel phase. This test can be used to segregate the bad emulsions from the good. It can also give an idea of the rate of deterioration if measurements of the same sample are made over time.

Emulsion explosives' strengths are also measured by underwater energy measurements, crater method, and lead block expansion. Microscopic observation of the emulsion over a period of time also shows the growth rate, size, and number of air bubbles in a specific area in a chemically gassed matrix and the data can be used to know the condition of the emulsion and predict the remaining shelf life.

- *Safety tests*: These are performed on a matrix/explosive separately. Impact, friction, spark, and static sensitivity are measured using standardized tests.

DTA/TGA measurements give a good picture of decomposition temperature and the heat released. These values give a fair indication if any unusual hazards are inherent due to impurities, catalysts, or new substances interacting with oxidizer salts such as AN and SPC.

6.8
Explosive Properties of Emulsion Matrix/Explosives

In most cases the matrix of emulsion has nonexplosive properties and these need to be established by performing tests described already.

A nonexplosive matrix should show negative results for initiation. Sometimes a matrix which is nonexplosive when tested at ambient temperature can become an explosive when tested at higher temperatures, say 80 °C. Also the matrix can show greater sensitivity to initiation when confined in steel as compared to initiation in the open. In such instances appropriate safety measures meant for handling explosive substances need to be implemented. This is understandable as the matrix is nearly oxygen balanced and contains both oxidizer and fuel in the right proportion. The only sensitizing ingredients absent are microbubbles. Thus, a matrix when produced at density of 1.38 should not show any explosive activity as it is at a density level much higher than critical density. It is apparent that lowering of the density is necessary to convert a nonexplosive matrix to an emulsion explosive. The extensive contact area between the oxidizer in the form of dispersed microdroplets and the fuel present as a continuous phase enables the emulsion explosive to approach much closer to ideality in detonation than any other multicomponent explosive.

Therefore emulsion explosives follow all the postulations of ideal detonation and VOD increases with density up to its critical density before failure.

The VOD of the emulsion explosive increases with charge diameter before leveling off but the increase is much more gradual even from small diameters indicating that even in small diameters, the emulsion explosive shows greater approach to ideality. All explosive properties become better with increasing charge diameter.

The sensitivity versus density relationship is also clearly seen when the emulsion explosive is subjected to initiation tests at low temperatures. At density 1.20 g/cc for chemically gassed emulsion at room temperature (25 °C), initiation by a No. 6 strength detonator is positive. The same emulsion at −10 °C is non-cap-sensitive. However, for the same density if microballoons are used, it is possible to retain cap sensitivity even at −10 °C .This may probably be due to the fact that crystallization of AN in the dispersed phase at the extremely low temperature does not disturb the air bubble structure as they are enclosed within strong physical barriers in synthetic microballoons.

Hydrostatic pressure affects the performance of an emulsion explosive. Due to the compression of the air/gas bubbles at the bottom of the borehole explosive density increases. If it goes beyond critical density under confinement for that diameter, failure to initiate/propagate can occur. The same phenomenon is avoided to a great extent by the presence of small amount of synthetic microballoon bubbles. These are strong enough not to get distorted or move under pressure and have the capability to remain as discrete hot spots and sustain propagation.

While discussing the fundamentals of detonation theory it is mentioned that use of detonation catalyst tends to speed up the reaction front velocity and lowers the decomposition temperature of the main ingredient, namely AN. The explosive then shows greater initiation sensitivity. The emulsion explosives being closer to ideality are well suited to test the effect of detonation catalysts. It has been observed by Wade and quoted in his patent USP 3765964 that strontium ion functioned as a detonation catalyst. The addition of 2.0% strontium nitrate enabled sensitivity to be increased whereby cap sensitivity was achieved even in 1 in. diameter cartridges with a No. 6 detonator.

The explosives properties of an emulsion explosive are influenced by the presence of sensitizers such as perchlorates, MMAN, and water [12] which also play an important role in providing an effective means of holding the oxidizer salts in solution. Thus, greater the water content lower the crystallization point of OB and easier it is to emulsify. If a sufficient quantity of the correct emulsifier is used, a strong emulsion can be obtained. However, increased water content lowers the strength of the explosive.

Table 6.6 VOD of different explosives confined and unconfined.

Product	VOD (m/s) unconfined 4 in. diameter	VOD (m/s) confined 4 in. diameter
AN/FO	4500	5000
Water gel	4800	5400
Emulsion	5500	5800

The addition of AL fine powder increases the strength, density, and sensitivity to detonation. The VOD decreases with increasing AL content due to after burning and endothermic reaction of formation of AL_2O_3. Thus beyond 4% Al does not bring benefits commensurate with cost. The use of flake AL powder in emulsions (contrary to water gels) did not bring any greater sensitivity than atomized powder and in fact appeared to lose its burning speed due to desensitization by the oil/wax in the fuel phase. The use of AL flake powder up to 5% even in conjunction with aeration (without use of microballoons) did not impart cap sensitivity in emulsion explosives.

The effect of confinement on the VOD of emulsion explosive is much less due its nearness to ideal detonation than for AN/FO. This is shown in Table 6.6.

The effect of temperature on sensitivity to initiation and propagation is much less for emulsion explosives than other types of explosives since even at $-15\,°C$ the inorganic oxidizer salts exist as a supersaturated solution in the dispersed droplets.

6.8.1
Channel Effect

A puzzling phenomenon was reported during underground blasting of emulsion explosives in small diameters. Blast failures and propagation failures were noticed even though there were no apparent causes of failure. Even when the explosive was found sensitive and propagated when tested in the open, some failures were noticed in actual boreholes after initiation. These results were more pronounced in earlier products using thick walled glass microballoons as sensitizers. The explosives when recovered and reinitiated failed to detonate again. Their densities when checked were found to have gone up beyond the critical density. It appeared that desensitization could be due to pressure from a shock wave generated in the adjacent borehole which had detonated earlier due to use of delay detonators for initiation. Later products of hollow spheres used in emulsion explosives as sensitizers avoided this phenomenon by having a thin wall resilient to pressure which sprung back to normal after the passage of the shock wave. Thus the explosive density was restored and so was its detonation sensitivity.

In another postulation, the phenomenon was ascribed to the channel effect. This was noticed when the diameter of the borehole was substantially higher than that of the cartridge used and there was a definite annular space between the explosive and the walls of the borehole. The channel effect was described to have taken place when the precursor shock wave (PSW) moves ahead of the detonation wave in air found in the annular space, precompresses, and desensitizes the unreacted explosive charge. This can cause detonation failure. Experimental work carried out using high-speed photography demonstrated that reduction in the PSW detonation velocity reduced the incidence of channel effect occurrences. It was also observed that the roughness of the inner surface of the borehole could reduce the PSW velocity. The experimental results indicated that the increased roughness tended to reduce detonation failure.

Vermiculite is a natural mineral which expands when heated. The process of expansion is called *exfoliation*. Vermiculite contains mostly oxides of silicon (as SiO_2), AL (as Al_2O_3), magnesium (as MgO), and iron (as Fe_2O_3). Exfoliated product has a bulk density of $4-10$ lb/ft^3. Specific heat $= 1.08$ kJ/kg and sizes vary from 16 to 0.5 mm of each particle.

Vermiculite is used in its many forms as filler, water retention agent in concrete. It has a honeycomb structure but is open on the sides. The structure is not totally enclosed, and hence while there are air pockets, they can be displaced.

Vermiculite after exfoliation was tried out as sensitizer in slurries and emulsion but without success. While it stabilized the density, it in itself did not contribute to sensitivity as it did not possess an enclosed air holding microstructure necessary to function as hot spots and be beneficial to an emulsion matrix in its explosive properties. At best it can be used in small percentages as a harmless filler/bulking agent in case explosives with less strength are needed but without reduction in density.

Perlite is natural silicious product with internal moisture and expands when heated for a long time with its original volume reaching the bulk density of $30-150$ kg/m^3. Basically it consists of SiO_2 and AL_2O_3 and has been tried out in an emulsion explosive as a sensitizer.

- *Fly ash*: Ash collected in the smoke stacks of thermal power plants also contains micro hollow spheres formed during the combustion of coal. However, the number of such spheres in a given volume of fly ash is rather low. Fly ash microspheres after segregation from bulk fly ash did prove to impart sensitivity to an emulsion matrix but in commercial scale it has not been separated, and hence hollow spheres from fly ash are not available in large volumes.

6.9
Permissible Emulsions

In spite of the fact that use of explosives for winning coal in underground mines has not seen any surge in demand, still the subject of efficient and safe blasting has been active. While in the early days permissible explosives containing NG were dominant, in recent times initially water gels and currently emulsion explosives have been predominantly used. It was thought that when slurry explosives were developed for bulk usage and for large diameter, it would not be technically easy to make permissible explosives with AL as a sensitizer. But to everyone's surprise AL flake powder-based water gels with varying NaCl and H_2O content were able to pass even the highest category (P5 class) test requirements applicable as per SMRE, Buxton, UK procedures which are mostly used as standard when describing permissibility ratings.

When emulsion explosives were discovered as another alternate to NG-based explosives and to AN/FO, it was once again not thought to be easy to formulate and manufacture permissible emulsions. Events have proved otherwise and to-day permissible emulsions cover most of the safety requirements laid down by statutory authorities for underground blasting in gaseous coal mines. The high VOD exhibited by emulsions even in diameters as low as 1 in. (25 mm) was a factor considered to act against passing permissibility tests especially those where direct initiation was used. Near ideal detonation had also put up the explosion temperatures beyond the limits normally considered as the limit for permissible explosives. Still it has been practically possible to establish permissible emulsions which have passed gallery tests creditably and these are being manufactured in large quantities and put to use with no adverse effect.

The design of permissible emulsions is based on the facts that water and chloride ions act as coolants and flame quenchers. Unlike in water gel permissible explosives, the NaCl which essentially supplies the chloride ion is present both in solution and as solid particles.

The NaCl as solid fine powder added into finished emulsion acts in a manner similar to the course/fine salt added in NG-based permissible explosives for performing as coolant.

A typical permissible emulsion could have a formula shown in Table 6.7.

Apart from NaCl, ammonium salts of oxalic, ammonium bicarbonate, ammonium chloride, and ammonium fluoride can also provide suppressing action against flame propagation and help the emulsion explosive to be safe in usage against ignition of gas/air/coal dust mixtures. The fact that AL is absent in the explosive composition is an advantage while designing the product for permissibility. Chemical gassing is also not a hindrance against permissibility. The effect of hollow

Table 6.7 Composition of typical permitted emulsion explosive.

Ingredients	Percentage in P1 type permitted explosive	Percentage in P3 type permitted explosive
Ammonium nitrate/ sodium nitrate/ calcium nitrate	65–75	60.0–70.0
Sodium chloride	5.0–8.0	10.0–11.0
Water	10.0–12.0	12.0–15.0
Emulsifier	1.2	1.2
Oil/wax	6.0	6.0

Table 6.8 Lead block expansion values for permitted explosives (ml) and density (g/cc).

Permissibility rating	Emulsion explosives	NG explosives	Water gels
P1	240–260 (1.15)	250–270 (1.30)	240 (1.15)
P3	210–225 (1.10)	220–230 (1.30)	200 (1.15)
P5	155–165 (1.10)	140–160 (1.20)	150 (1.15)

microballoons on permissibility has not been fully understood or investigated in detail. While their function as effective hotspots and triggers for detonation is known, their influence on initiation of methane air mixtures is still to be explained. Emulsion explosive composition with 1% of microballoons has been able to pass P1 and P3 category permissible ratings. Higher microballoon percentages at 3% have caused problems in incendivity tests for higher categories (P3 and P5) of permissibility specified for explosives to be used in highly gassy underground coal mines.

Strength measurements using expansion in lead blocks have shown gradually decreasing values as the category rating increased. Table 6.8 shows the obtained values for three different types of explosives.

Permissible compositions with sodium perchlorate and $Ca(NO_3)_2$ could function at higher densities. In all the three permissible categories the corresponding NG explosive had higher densities but in the P5 category the lead block expansion of an emulsion explosive was higher due probably to the larger gas volume and higher VOD.

6.10
General Purpose Small-Diameter (GPSD) Emulsion Explosives

The true test of any GPSD explosive is its ability to pull hard rock in tunnels. NG explosives (Gelatins) with their high density of 1.4 have set the benchmark and it was a challenge to the formula designer to come up with an equally high performing emulsion explosive. This objective was not achieved in the initial stages due to inherent sensitivity problems at higher densities. Such formulas were operated at 12% water content. Use of very high percentage of AL was limited due to cost and cost benefit factors not being favorable. The only way of getting the density higher than the 1.15 g/cc level which was the norm then was by lowering the water content. With better emulsifiers (blends), mixed gassing systems, and use of additional sensitizers such as MMAN or sodium perchlorate, it has been possible to increase the density of the product to 1.25–1.30 g/cc with water content between 8 and 10%. It has now been reported that even a lower water content of 6% has been used to reach the product density of 1.35 g/cc. It is obvious that while water content has gone down the oxidizer salt content has gone up, and to obtain a supersaturated

Table 6.9 Composition of typical GPSD (no microballoons).

Ingredients	Percentage in composition
Ammonium nitrate	73.50
Sodium nitrate	5.0
Sodium perchlorate	4.0
Water	9.0
Complexing agent	0.20
Surfactant	1.0
Vaseline	1.0
Paraffin wax	1.80
Microcrystalline wax	0.60
Emulsifier blend	2.40
Long chain fatty alcohol	0.50
Gassing agent	1.0

oxidizer solution processing temperatures in excess of 90 °C are needed even if part of the AN is added later into the emulsion as prills or as AN/FO. This means that more energy is needed to be spent to cater to the extra steam requirement for operating at this temperature. Wax content also needs to be increased so that fudge point does not get too far below the OB crystallization point. This may require use of higher content of microcrystalline wax adding to costs. Due to high oxidizer salt content there will be massive crystallization occurring in the dispersed phase as the emulsion cools. These crystals mostly of AN have to be sensitized either by cocrystallization with other salts such as $Ca(NO_3)_2$, SPC, or MMAN or by adding crystal habit modifiers such as SLS so that the crystals coming out of the supersaturated solution will retain sensitivity. Detonation catalysts may have to be used to increase the sensitivity to initiation. Notwithstanding all these difficulties formulations are available to produce GPSD emulsions of high strength which are also capable of competing with NG-based explosives in most applications. In particular, much research work done earlier in Sweden and China has resulted in commercialization of such products. The flip side is that the product with the low H_2O content tends to become closer in its reaction to external stimulus to NG-based gelatins, and thus the design of the process/plant/handling will have to accommodate such characteristics.

A typical formula which gives a GPSD product that can be packed in paper cartridges like dynamites at density 1.20 g/cc, lead block expansion at 330 ml, and VOD at 3750 m/s in 1 in. diameter in the open initiated by the No. 6 detonator is given in Table 6.9.

The product comes out as a thick emulsion and solidifies on cooling. In spite of the high AN and low H_2O contents, the product remains without massive crystallization for four to six months in tropical storage with weeks of cycling around 32 °C. Significant is the high emulsifier content which gives a lot of reduction to the interfacial tension.

This explosive can detonate quite well even at −5 °C in spite of the absence of microballoons. Variations of the above formulas exist to suit RM availability and cost. However, this product cannot be pumped and hence cannot be packed in a chub machine to produce sausage-type poly film cartridges. Extruding-type packing machines are used to push the thick emulsion into preformed paper cartridges of desired diameter and length. The productivity is lower than high-speed chub machines.

A formula which is more suitable to pump and pack in a chub machine has been made with slightly lesser strength but with greater productivity in packaging. The product showed a density of 1.15–1.18 g/cc, VOD of 3800 m/s in 1 in. diameter in the open, and lead block expansion of 280 ml. The crystallization point was 69 °C.

In case microballoons are preferred to be used partially instead of total chemical gassing, the matrix needs to be made less viscous to accommodate the microballoons without formation of a stiff matrix difficult to pump and pack. Then a formula with more oil and less wax is used. Such an emulsion explosive had a crystallization point of 65 °C, a VOD of 4000 m/s in 1 in. diameter in the open at density 1.20 g/cc, and a shelf life of six months minimum. The oxygen balance was −3.0.

In case it is desired to lower the crystallization point further for easier processing and reduce expenditure on steam, part of the AN is added after forming the matrix in the form of AN prills. It is necessary to add sufficient fuel in the matrix to more than balance the AN prills to be added later on. The crystallization point came down to 60 °C and the explosive gave a VOD of 3600 m/s in 1 in. diameter in the open at a density of 1.15 g/cc. The explosive also showed a lead block expansion of 270 ml and a shelf life of four months minimum.

Detailed composition of all these various compositions is shown in Table 6.10.

With all the GPSD formulations because of the low water and high oxidizer salt content, great care needs to be taken to avoid excessive friction and impact all through the process and also prevent drying out of the emulsion as this increases the hazard level substantially. The number of patents on formulation and process concerning emulsion explosives is huge but more important pioneering some relevant ones [10, 12–19] show interesting findings.

6.11
Bulk Emulsions

By far the greatest volume of emulsion explosive is used as a bulk explosive for direct pumping into boreholes. The options available are to produce the explosive emulsion fully finished (except for the addition of AN prills or AL powder) at a base plant, take it to the mine site in a bulk delivery truck, and pump into the boreholes. Before delivering into the boreholes, the required quantity of the solid ingredients is added. In this method emulsion is made and handled as an explosive right from the mixing stage in the base plant. Hence, the base plant needs to observe all the

Table 6.10 Composition of different emulsion explosives (GPSD).

Ingredients	Percentage in product for chub style packing	Percentage in product with microballoons	Percentage in product with AN prills addition	Percentage in product with MMAN
Ammonium nitrate	61.0	60.0	51.0	45.0
Sodium nitrate	6.0	8.0	5.0	—
Calcium nitrate	8.0	8.0	8.0	8.0
Sodium perchlorate	5.0	4.0	5.0	5.0
Mono methyl amine nitrate	—	—	—	25.0
Water	11.0	11.0	11.0	11.0
Waxes (paraffin and microcrystalline)	2.5	2.0	2.0	1.5
Microballoons	—	1.0	—	1.0
White mineral oil	3.5	3.2	4.0	1.5
Atomized Al	—	—	0.5	—
Emulsifiers	2.0	2.3	2.0	2.0
Gassing agent	1.0	0.5	0.5	0.3
AN porous prills	—	—	11.0	—

statutory requirements for safety distances, man, and explosive limits which may render setting up such a plant a costly proposition.

The emulsion explosive can also be produced on a truck starting from OB and F/B but the quantity produced per truck is limited by RM logistics. The production truck is really a mini manufacturing plant and is expensive for its production capacity since the carrying capacity is spread over all the RMs.

The most popular and sensible option widely practiced today is to produce a nonexplosive emulsion matrix in the base plant and load it onto a delivery truck which has provision to add/mix liquid and solid ingredients before delivery. Such trucks can carry up to 20 tonnes of matrix and convert it to explosives for delivery. These pump trucks can be fed continuously on the mine face by tankers containing the nonexplosive matrix. By making a nonexplosive matrix only at the base plant, the land and other measures exclusive for an explosives manufacturing facility can be avoided. The nonexplosive matrix as the name suggests is a chemical mixture of oxidizer and fuel but has no explosive properties. It cannot be initiated by even a booster at the density it is made and kept. A shelf life of two to three weeks is more than sufficient. Usually a typical matrix composition (nonexplosive) is

OB composition (%)	FB composition (%)	Matrix composition (%)
AN 67	HSD 28	OB 93
CN 8	FO 56	FB 7
SN 3	SMO 16	
H_2O 22		

Matrix density is kept at not less than 1.36 g/cc.

To 100 parts of a nonexplosive matrix 0.25% gassing agent (10% sodium nitrite solution) is added and thereafter 10% AN prills is mixed into the matrix before delivering the product into the borehole. The final composition will be as given below:

H_2O	18.6	Density	1.15 g/cc
SN	2.5	VOD	5000 m/s in 4 in. diameter in the open when freshly made and boostered with Pentolite
CN	6.7		
AN (total)	65.0		
Fuel	5.4		
Emulsifier	1.0		

When AN prills are added to the emulsion matrix, the product is also called "Doped" emulsions. Due to the reduction in H_2O content and increased AN content the product is expected to have higher bulk/weight strength and more gas volume generation. In most bulk formulations such as that given above due to high water content and a loose emulsion formed due to low energy mixing, even after using a gassing agent and Porons AN prills, the product is only booster sensitive and that too in 4 in. diameter and above. The advantage of this product versus heavy ammonium nitrate fuel oil (HANFO) is that it can be pumped at good loading rates (300–400 kg/ min) as the product is still preserving the viscosity characteristics of a flowy emulsion and it sinks in watery holes to the bottom because of its higher density than water unlike AN/FO.

6.12
Heavy AN/FO

There is always a doubt whether this product should come under the AN/FO category or as an emulsion type. I have included it here as a part of emulsion.

The reason being that the emulsion plays a crucial part in improving some of the characteristic of AN/FO not particularly desirable. Use of emulsion does the following to AN/FO.

1) Raise density
2) Raise water resistance
3) Raise VOD
4) Raise sensitivity to boostering
5) Raise shock wave intensity.

HANFO came into being after pure emulsions were established and were available as an alternative at (slightly) higher cost but with a lot of additional benefits for bulk loading and blasting. Under some conditions using pure emulsions may not be necessary but use of ANFO may not be fully adequate for optimum blasting results. As a via media, mix of emulsions and AN/FO was considered and found beneficial. The usage of HANFO surged initially to a great volume but in recent years pure emulsions have become standard product for bulk loading. The reasons could be twofold.

1) For best results, an emulsion already finished into an explosive needs to be used as one of the components, and hence availability of a pure emulsion is mandatory.
2) If an emulsion explosive is to be made from a nonexplosive matrix first on the truck before mixing with AN/FO, then the truck becomes more complicated and cost increases.
3) There is need to make AN/FO separately to be added later. The second component AN/FO can also be made on the bulk truck before it is mixed with the emulsion explosive. This is much easier than that in [1]. This concept of making AN/FO on the truck and blending it with finished emulsions is practiced quite a bit in the USA and Australia. The proportion used is 70/30 ANFO/emulsion. The mix is dry and is augured and used in dry and moist holes of very large diameter (7 in. and above).

Proportion of 50/50 (AN/FO to emulsion explosive) gives a product with greater water resistance but still not of pumpable viscosity. The system of HANFO allows a lot of flexibility in the proportion of AN/FO to emulsion explosive and this can be utilized optimally depending on the blast site-specific requirements. The properties of emulsion/AN FO/HANFO are shown in Table 6.11.

The benefits derived by HANFO is due to the emulsion explosive filling the empty space between the spherical prills of AN/FO particles.

This increases density and VOD. The AN/FO prills are now coated with a layer of emulsion and hence develop water resistance also and performance becomes much better than regular AN/FO.

In many instances for ease of operations and use of a simpler bulk truck, "doping" with AN is resorted to rather than delivering HANFO. Even if oxygen balance is maintained for the extra AN added by taking more fuel in the emulsion matrix, it is not equivalent to HANFO where a good well mixed and balanced

Table 6.11 Comparison of emulsion explosives with An/FO and HANFO.

Properties	Emulsion	AN/FO	HANFO (30/70)
Density (g/cc)	1.15–1.20	0.95–1.00	1.05–1.10
VOD (4 in. diameter open) (m/s)	5000–5200	4000–4300	4500–4600
Weight strength	120	100	110

AN/FO prepared separately is mixed with the emulsion. Here there is no doubt that the entire mix is oxygen balanced while there is a chance in case of doping that AN combustion in the presence of fuel is not ideal and can lead to loss of performance and generation of toxic fumes. The authors' recommendation is always to use HANFO rather than a "doped" product.

6.13
Packaged Large-Diameter Emulsion Explosives

Although bulk emulsions are produced in very large volumes, still there is sufficient demand for packaged products in medium diameters from 60 to 125 mm. These are used in watery holes such as deep coal seams and in smaller open cast mining such as limestone, in large canal construction, and quarries for road aggregates. In these applications where a borehole cannot be loaded more than 100–150 kg on an average, pumping of emulsion becomes impracticable due to the high pumping rate. Even at 200 kg/ min, the hole is filled up in half a minute, hardly enough time to insert the hose down and remove it. The total explosive per blast is also small and much below optimum capacity utilization for a bulk truck.

Packaged emulsions are formulated with a good strong base emulsion to withstand leaching by water in case the package breaks in the borehole as it does sometimes when dropped into deep holes. The product also needs to withstand pressure due to a long column head in deep holes (coal seams). The density should be sufficiently high that the product sinks even in muddy water. A typical booster-sensitive formula will have both chemical gassing and small amount of microballoons to ensure adequate sensitivity. A typical packaged product formula is given in Table 6.12.

An explosive with the above composition could detonate with VOD of 5000 m/s in 4 in. diameter in the open with a booster at density 1.20 g/cc after six-month storage.

For under water blasting, seismic applications, the product instead of being packed in flexible packing is poured into rigid couplable tubes. To ensure no failure at greater than 30 m depths the product needs additional sensitivity and hence will contain a higher percentage of microballoons or an additional sensitizer either sodium perchlorate or MMAN.

Table 6.12   Composition of large diameter packaged emulsion explosives.

Ingredient	Percentage in composition
Ammonium nitrate	66.0
Sodium nitrate	5.0
Calcium nitrate	8.0
Water	13.0
Paraffin wax	2.0
Emulsifier	1.2
Furnace oil	3.0
Microballoons	0.7
Coated atomized aluminum powder	0.7
Gassing agent	0.3

For a packaged product, much longer shelf life is expected than for a bulk product since it may have to be dispatched to faraway places from the base plant and may have to spend considerable time in storage magazines before use. Thus, the emulsion matrix needs to be strong and stable. Proper selection of emulsifier, use of stabilizer, and processing in a high shear mixer is necessary. The type of mixer used for producing bulk explosives is not adequate to produce emulsions of good long-term stability.

References

1. Wade, C.G. (1978) Proceedings of the 4th Conference on Explosives and Blasting Techniques, USA, pp. 222–233.
2. Wang, X. (1994) *Emulsion Explosives*, Metallurgical Industry Press, Beijing.
3. Griffin, W.C. (1945) US Patent 2,380,166.
4. Griffin, W.C. (1949) *J. Soc. Cosmet. Chem.*, **1**, 311.
5. Griffin, W.C. (1954) *J. Soc. Cosmet. Chem.*, **5**, 249.
6. Tadros, F. (ed.) (2009) *Emulsion Science and Technology*, Wiley-VCH Verlag GmbH, Weinhiem.
7. (1994) Uses of Lecithin. Chemical Weekly.
8. Becher, P. (1965) *Emulsion Theory and Practise*, Chemical Rubber Co, Scientific Press, Cleveland, OH.
9. Sherman, P. (1963) *Rheology of Emulsions*, Pergamon Press, Oxford.
10. Alchemie Research (1996) ICT 27th Annual Conference, Germany.
11. Schwarz, N. and Bezemer, C. (1956) *Kolloid-Z.*, **1–3**, 139.
12. Catermole, G.R. (1970) US Patent 3,674,578.
13. CIL Inc. Canada IP Comp Spec 16289.
14. Bluhm, H.F. (1969) US Patent 3,447,978.
15. Wade, C.G. (1973) US Patent 3,715,247.
16. Wade, C.G. (1973) US Patent 3,765,964.
17. Wade, C.G. (1979) US Patent 4,149,916.
18. Wade, C.G. (1973) US Patent ,318,281.
19. Tomic, E.A. (1973) US Patent 3,770,522.

Further Reading

Boyd, J., Parkinson, C., and Sherman, P. (1972) *J. Colloid Interface Sci.*, **41** (1), 359–370.

E I Dupont De Nemours Ind. Patent 154,470.

Ceshanski (1985) US Patent 4,496,405.

ICI India Ind. Patent 168,957.

MSI (1991) WO Patent 91/01,800.

Peng (1991) US Patent 4,992,118.

SASOL (1989) EU Patent 0,340,980.

7
Research and Development

An important expectation from writing this book is the hope that research will be done in certain gray areas of explosives' behavior. In many instances enough work has not been generated in order to explain some findings and observations thrown up while manufacturing and using explosives. It would add much to explosive sciences credibility if more explanations can be offered based on scientific findings. A list of such areas of work is made out to the best of author's knowledge. It is possible that more work has been done in these areas in recent times. Any communication to the author in this regard is more than welcome.

It is a fact though that there is hardly any time or money being spent in R&D work in the field of civil explosives. It is clear that in recent years the major work done is about development of formulations, evaluating safety characteristics, design of production processes and plants, rather than basic or background research specifically oriented toward science of explosives. More often concepts on detonation and thermochemistry postulated in the early 1950s have been used for explanations of phenomenon related to its behavior seen practically in the field while using AN-based explosives.

The nitroglycerine (NG)-based explosives are able to detonate to their full potential at higher densities than ammonium nitrate (AN)-based explosives. This could be as much as 40% more when compared to AN/FO (fuel oil) and 20% more when compared to slurries and emulsions. This reflects in the amount of explosives that can be loaded into a borehole. A loading factor of approximately of the same degree as the relative difference in density results in concentration of greater energy when NG explosives are used. The effect of this while noticeable in large-diameter blasting is much more pronounced when blasting in tunnels or in hard rock strata employing small-diameter holes. On the other hand the velocity of detonation (VOD) for NG-based explosives are consistently lower by 15–20% when compared to the water gels/emulsions both in large-diameter holes and even in small-diameter holes where explosive cartridges of diameter as low as 1 in. are used. The higher VOD leads to higher shock energy component which helps in fragmentation of the surroundings. To some extent the difference in energy density between AN- and NG-based explosives is compensated by the higher shock energy. When it comes to AN/FO, the higher gas volume generated per unit weight of explosive by AN/FO does help in a better performance and

Ammonium Nitrate Explosives for Civil Applications: Slurries, Emulsions and Ammonium Nitrate Fuel Oils, First Edition. E.G. Mahadevan.
© 2013 Wiley-VCH Verlag GmbH & Co. KGaA. Published 2013 by Wiley-VCH Verlag GmbH & Co. KGaA.

narrows the gap. But it can be seen that there is still scope for improvement in the AN-based explosives performance in order to meet and exceed that of NG explosives which have been considered as benchmarks by the end user for the past so many decades. Hence research in the past and probably in the future would still be oriented toward findings that will help the explosive industry to manufacture products with still higher energy outputs but at the same time safe enough to manufacture even if it means changing the model existing today drastically.

7.1
Areas of Interest

1) Air-gap sensitivity – (what factors really affect the air-gap sensitivity?). influence of explosion temperature, size, and nature of metallic particles, self-explosive ingredients, and effect of confinement on air-gap sensitivity.
2) Study of behavior of some compositions when subjected to pressure (performance in medium diameters in deep borehole blasting).
3) Replacement of water in part or in full by other liquids in the water gels and emulsions to enhance density and cold temperature sensitivity.
4) Improvement and utilization of modern technology in safety aspects of explosives manufacture especially early warning systems and other techniques like venting and suppression.
5) Phenomena of cross-linking as means of attaining gel stability (different types of gums and different cross-linking).
6) More efficient fuels (water-soluble/water-miscible). Is the AN crystallizing out of the gel having enough fuel to function efficiently?
7) Estimating the amounts of AN in solution and out at various temperature of a water gel and how to alter this ratio for better explosives performance.
8) Role of water, NaCl, and other coolants in permissibility.
9) Developing a laboratory scale measure for permissibility. Gallery testing is cumbersome/expensive. Can a mini gallery type testing be developed for screening various formulations quickly and easily?
10) Mechanism and theory behind the ability of certain compounds to promote permissibility.
11) Study of compounds as possible replacement (part or full) of AN.
12) Study of ways and means to increase the density of finished product (explosive) without loss of sensitivity.
13) New compounds to enable use at very high and low temperature.
14) Micro emulsions – for use as explosives emulsion.
15) Study of polymeric systems for use in formation of emulsion (explosive) matrix.
16) More two component systems.
17) Generation of finer gas bubbles and stabilizing them.

18) Developing a mini underwater test facility which can be easily set up and operated.
19) Investigation of greases or oil in water (O/W) emulsions for explosives, advantages being easier to make, wider choice of emulsions, easier permissibility, better stability, and use of polymer system in aqueous phase.
20) Study of the effect of raw materials on the performance characteristics of explosives; the raw materials could be metallic or nonmetallic. Of interest would be the effect of nanoparticles in explosives performance (nano AL, nano Fe or Fe_2O_3, sulfur, microballoons, and AN).

7.2
Development Work and Upscaling

Development work in explosives involves first working in the laboratory with small quantities, standardizing the conditions of preparation and formulation, then moving on to pilot plant scale operation repeating the laboratory work in a larger scale, and then finally moving to plant-scale operations. Even here the plant batches are progressively increased in order to avoid failures and being saddled with defective explosives. The laboratory scale for making batches of slurry/emulsions/AN/FO are usually 1–2 kg or occasionally 5 kg. Every extra quantity of explosives made will place severe strain on man and safety distance limits. Most of author's work was done on a 1–2-kg scale. Laboratory batch making requires a good mixer (usually a vertical variable speed clover leaf mixer) with flame/explosion-proof electrical motors and fittings. The amount of explosive is small enough to pack by hand. Usually laboratory batches are made to develop small-diameter, cap-sensitive products. For large-diameter product development, pilot plant batches need to be made because of increased quantity of explosives needed for testing. The pilot plant batches can vary from 5 to 50 kg of explosive quantity per batch with different capacity and type of mixers being used. Double helical horizontal type, vertical planetary type, continuous mixers should all be of matching capacities. Generally batch size is around 70% of mixer volume. Necessary heating/cooling facilities, hopper for large diameter packing, individual hand packers for small-diameter packing, and clipping machine are needed.

It is author's experience that at least 10 laboratory batches, 5 pilot plant batches, and 20 plant batches are required to establish a product for field trails. At least 100 tons of explosives of the same formula are required to obtain reliable feedback on field performance for large-diameter packaged product and about 1–2 tons for small-diameter product. The time period can be 4–6 months from start of laboratory batch as at least 3 months of normal storage data is needed, even though accelerated storage can give an idea of product stability much faster. This data can be used for quick selection of formulas to go further. Performance and stability tests are conducted on the batches at regular intervals to obtain relevant data.

Thus much advanced and meticulous planning is needed from the start of a product development program. Factorial design of experiments should be used to reduce the number of experiments needed to obtain the required data.

Author's experience is that pilot plant should be located at a different location than the manufacturing plant to avoid the tendency of going into plant trials without adequate data from pilot plant operation. A different location could also help in preserving confidential information.

The operational and safety procedures, housekeeping, and type of personal involved in R&D should all be of a level not less than those practiced in the plant. In fact since new materials, processes, and process conditions are tried out very often, greater vigilance on following of safety rules is needed such as wearing of safety goggles for eye protection, two men present at all times, presence of adequate firefighting equipment, instructions on how to deal with fires caused by different types of substances, and display of instructions for operation of equipment especially pumps/mixers.

More accidents or near accidents have been caused in pilot plant operations than in the regular manufacturing plants though because of smaller quantity of explosives involved the effect could be less damaging.

Instrumentation allied to pilot plant operations are (i) viscometer, (ii) pH meter, (iii) safe thermometers, (iv) balances, and (v) test apparatus for density measurements.

For stability evaluations, ovens to house at least 5–10 kg of explosives are needed. These ovens can be steam or hot water heated to avoid electrical connections, with temperature controllers and cut offs. The oven should be housed in a regular barricaded magazine-type building.

A freezer with a few kilograms of capacity needs to be housed similarly. These are needed to conduct freezing/thawing tests and cold temperature sensitivity studies.

A firing site for testing of explosives is an absolute must. An underwater test facility would be useful but if not nearby, other methods of strength estimation such as lead block expansion, crater test needs to be performed and facilities for these need to be available.

One should not ignore packaging since considerable quantity of AN-based explosives are still transported and sold in small packages (20–25 kg in weight) and in cartridges of small lengths and diameters, in paper and plastic shells. Hence suitable equipment needs to be available to carry out tests on packing material, the results of which could decide their usage in trials initially and if successful in large scale. Test equipment for determining bursting strength, adhesion, elongation, penetration strength, porosity, and permeation to water vapor are needed. Drop tests on the final package are also necessary to determine the ability of the package to withstand transportation and handling. The need for elaborate testing of packaging material to arrive at its suitability is guided by the various international codes and specification of packaging material statutorily imposed due to the hazardous nature of the goods being transported. No doubt the need for such elaborate search has come down in recent years as greater standardization have occurred and also the use of new designs and materials are considered only if there is a cost advantage to the existing which has proved to be adequate for most situations of transport.

Meticulous writing of log book on batches made with full details of process used, conditions, and observations need to be kept. The same batches need to be followed up with performance and stability tests initially and later field performance reports have to be obtained and correlated with laboratory and pilot plant data. All in all, it is a comprehensive challenging assignment and gives much satisfaction on completion of a project. In most cases, the ultimate success of either new or modifying the existing formula, process, equipment, and packaging lies in its acceptance, and introduction on a regular basis at the manufacturing plant. A (joint) program of handing over between R&D and plant is most useful and is a key to success of R&D programs. Collection of safety data starting from laboratory batches to pilot plant operations should lead to safety procedures for handling and storage of all newly developed products and raw materials used and this information should be handed over at the right time to plant personnel along with a manual for production.

The manual developed by R&D initially, modified finally together with production personnel if necessary, should be complete. Apart from describing the process fully with all operating conditions, raw material checks, in-process/final quality checks, product specification and release tests, and detailed safety guidelines should be included. Minimum number of copies should be kept with authorized personnel. All the above requirements appear to be what should be routine or normal in any good manufacturing practice but surprisingly it is very often absent or distorted and hence their importance is once again stressed.

7.3
Management of R&D

The experiences of my years in R&D in the explosive industry in various commercial explosives manufacturing organizations have convinced me that while there are similarities with other areas of industrial R&D, there are some unique features also which one needs to be aware of. These can be highlighted as follows:

- Difficulty in getting ready-made scientists/engineers in this area of work
- Difficulty in coming out with quick solutions since they require long time for establishing as viable solutions.
- The need to be always careful of safety matters.
- The need to be always conscious of costs involved.
- The need to work within regulatory boundaries all the time.
- The need to restrict publications for reasons of secrecy.
- The lack of contact with like-minded workers for free communication and problem solving.
- The need to get the findings accepted and established in manufacturing and end use where people of different education, background, and objectives rule.

Every organization may develop different models to successfully establish and run an effective R&D keeping in mind the above unique features inherent in the area of civil explosives. Few have succeeded in coming out continuously with new innovations because of the limitations. The rate of totally new revolutionary product established commercially on a sound footing is low. History shows it to be almost one in a decade. Rate of obsolescence is also very low. However refining and upgrading of an established product is a continuous activity and forms the major part of the R&D work being carried on and published mostly in the form of patents.

It is the author's view based on his experience that the R&D organizational structure should be such that it facilitates frictionless transfer of R&D findings to commercial success which means going through plant and field trials. A good way of doing this without exposing the scientist/engineer directly to the outside would be to use an R&D coordinator who will act as the intermediary/liaison between R&D and outside departments. The R&D coordinator also takes care of all administrative responsibilities of running the R&D department and leaves the technical personnel free to concentrate on their projects.

8
Functional Safety during Manufacture of AN Explosives

8.1
Introduction – Personal View Point on Safety

In the world of explosives, the word safe (ty) is frequently used sometimes wrongly. To me there is no such thing as *safe*. It can only be *safer* than something already known [1].

Nitroglycerine (NG) explosives were considered unsafe and difficult to handle in the initial stages of its development, later considered safe and manufactured and handled in large quantities, but in the last decade again considered less safe than other explosives made without NG as an ingredient. The philosophy thus boils down to a comparison between substances already in use with some new discoveries. It is now standard procedure to classify the hazardous nature of a material by its behavior toward energy input in the form of impact and heat. Tests have been developed based on friction, shock (thermal and mechanical) to judge, and estimate fairly accurately the behavior of substances under the influence of such stimuli. Knowledge gained on the sensitivity of explosive substances has been used to design, operate, and handle explosives in a safe manner.

However this does not mean we have arrived at a point of zero occurrence of accidents/incidents in the explosives industry. On the other hand, the information on such happenings over the last decade shows that there is still a long way to go. As said earlier one can only strive toward a perfect accident-free situation continuously and there is always scope to move toward something safer, whether it is manufacture of explosives or road safety.

Much has been written about safety in all aspect while handling hazardous materials including civil explosives. A recent book by Agrawal [2] gives a very complete picture of the different ways of inculcating good safety practices in manufacture, handling, storage, and transport of explosive substances. The same is also described in a very practical manner by Sreenivasan [3]. It is not the intention of the author to repeat these, but for greater impact specific sources of hazards identified over decades of experience in the explosives industry will be critically examined and guidelines for reducing the risk from such hazards discussed.

Basic safety philosophy involves commitment to the implementation of the core concept of safety in all aspects [4, 5]. In explosives industry the very nature of the

Ammonium Nitrate Explosives for Civil Applications: Slurries, Emulsions and Ammonium Nitrate Fuel Oils,
First Edition. E.G. Mahadevan.
© 2013 Wiley-VCH Verlag GmbH & Co. KGaA. Published 2013 by Wiley-VCH Verlag GmbH & Co. KGaA.

product being handled is hazardous and hence the path to be followed should be indeed the motto "prevent a hazardous situation before it becomes uncontrollable and where it goes beyond control minimize the damage to personnel." In order to implement this concept of safety, the emphasis should be on early detection of a potentially hazardous situation by means of modern detection equipment which will operate control measures without recourse to human input in order to reduce the dependency on human judgment of a hazard situation and speed up countermeasures.

In dealing with the business of safety in explosive manufacture and distribution, thumb rule for reducing risks and damage follow these tenets:

1) Reduce quantity of explosives handled in process.
2) Reduce the impact of energy input to the explosive by striving to operate at lower temperature, pressure, and kinetic energy input.
3) Reduce influence of external/internal sources of impact, thermal shock, friction, pressure, and static electricity.
4) Reduce exposure of human beings directly to explosives.
5) Reduce number of human beings in an explosive operation.
6) Use mounding/barriers frequently and effectively.
7) Use separation according to safety distances calculated as per the classification of the explosives being handled.
8) Do not ignore fire hazards and its influence on the rest of the operations/materials.
9) Do not ignore explosion hazards from other sources such as dust explosions, electrical motors and cables, and lightning.
10) Guard against sabotage and terror acts.
11) Insist on good housekeeping.
12) Do not accumulate waste or rejected material, either rework or destroy – do not delay.
13) Take precautions while destroying waste/rejected material. Expect explosions.
14) Do not ignore fume offs and sudden exotherms; they are warning signs.
15) Do not ignore unusual sounds, torque, and voltage fluctuations. Stop and investigate.
16) Do not use extra force to remove joints with stuck up material.
17) Use non-spark-proof tools at all times.
18) Wash up every time when plant is stopped for longer time.
19) Use mesh to prevent foreign bodies from entering critical areas.
20) Use good ventilation and easily accessible exits in all operating rooms in the plant.
21) Use lightning arrestors for building and earthing of all equipment where generation of static electricity is expected or known to occur. Wear conducting shoes or use bare feet. No synthetic clothing should be worn.
22) Keep the floor nonslippery.
23) Prevent possible runaway reactions triggered through chemical/metallic impurities.

24) Use explosion venting as much as possible to protect plant and personal.
25) Use explosion suppression techniques for advanced warning and prevention of explosions.

8.2
Safety Considerations in AN Explosives

8.2.1
In AN/FO

ANFO being the simplest two-component (or at best with three to four ingredients) explosive it is easy to understand the risks involved. The major component ammonium nitrate (AN) has been thoroughly evaluated for its sensitive properties, so is the diesel oil as fuel. The advent of AN prills which do not set themselves into a hard mass has eliminated to a great extent the need to chip and break the caked lumps of AN, a leading cause for disastrous explosions in the earlier days. However the low-density AN prills do have a higher sensitivity because of the air voids in them and should be handled carefully. Diesel oil is an inflammable liquid with flash point low enough to catch fire as vapor/air mixture easily.

Statistics have shown that in 90% of the accidental explosion of ANFO, fire has been a precursor and source of thermal energy high enough to trigger explosion of AN and AN/FO. The diesel storage above ground in drums is a source of fire which can be ignited by spark or another fire source. The diesel fire in turn can cause rapid thermal decomposition of AN or AN/FO which can lead to explosion. The confinement AN is subjected to within a body of stored AN is many times sufficient to provoke AN to explode and cause damage to the surroundings. If other types of explodable substances are stored nearby within the safety distances calculated on the mass of AN present, there is every chance of these going off causing greater destruction.

The ease with which AN can decompose into a runaway reaction which could result in an explosion is magnified by the porosity of the AN, chemical impurities like nitrite, decomposition catalysts the Na dichromate, and iron oxide powder which may be present as impurities in the AN.

While rough handling of AN is certainly to be avoided and nonsparking tools used, more critical is the fire safety measures that need to be adopted. The firefighting measures to be used once the fire has started in an AN/FO operation have already been well studied and guide lines given.

The basic principle of separating AN store from diesel oil must be practiced in all seriousness by segregation and fire propagation barriers.

An interesting finding has been that most of the incidents involving ANFO have occurred while transporting it, especially in the USA where transport of ANFO by road is done in huge volumes. Road accidents involving ANFO leading to explosion has occurred only after there has been a fire caused in the accident or when there has been a mixed cargo of other inflammable materials. Outside

USA, stationary plants manufacturing AN/FO have been involved in explosions once again after fire has been noticed. Usually a measurable delay has taken place between start of fire and subsequent explosion. Based on the difficulties encountered in putting out a fire after it has taken a foothold, evacuation to safe distance of personnel is highly recommended and the fire allowed to burn by itself till it is extinguished.

Thus fire warning systems, smoke detectors, video surveillance of storage area, firefighting equipment, presence of nearby fire station, and wind direction indicator are all steps absolutely necessary to have a reasonable comfort level of monitoring fire detection and fighting propagation. Basic concept of reducing stockpiling AN and FO is important. Just in just out method would ensure minimum quantity of vulnerable substances present in the area.

Effect of lightning is to cause a discharge of heavy voltage induced electric spark. It can cause a fire/explosion in inflammable substances and vapors. Protection against strike of lightning on buildings housing AN and FO is a must. Standard passive and newly developed active systems are available for installation.

Apart from protection against fire hazard, safety during manufacture of AN/FO is also in ensuring that screw mixer/conveyor do not get jammed or clogged and there are no hard metallic objects inducing friction as the AN/FO moves along.

8.2.2
In Slurries and Emulsions

Safety considerations in slurries/water gels/emulsions manufacturing is similar to that in AN/FO plant as far as handling and storage of AN is considered. There are, however, other areas of concern brought about by the use of other raw materials (RMs) in fine physical state or organic liquids and fuels.

The critical areas of hazard apart from the usual friction, impact and static concern dust explosions [6] and fire hazards involved in the use of fine reactive aluminum (AL) powder. The special ways of handling fine AL powder has already been described in this book earlier. Dust cloud formation can happen during production if fine materials like guar gums and sulfur are added open to the atmosphere. A mixture of guar gum, AL powder, and sulfur can form a potent combination in a dust cloud. There have been instances where such mixtures were sources of fire/explosion in a water gel production unit producing a cap-sensitive product. Starch is also well known as capable of forming explodable dust clouds. If a closed system for adding these ingredients is not used, blanketing with inert gas of the area where dust cloud is formed will be a primary step to prevent explosion. All dust explosions do need oxygen to sustain. Good ventilation itself is a great deterrent to dust explosions. Air currents dilute the concentration of the active ingredient in the dust cloud and reduce the risk of explosion drastically. Setting off dust cloud explosion is through a spark invariably generated in an electric motor/cable, impact of metal against metal or spark due to static electric discharge. Control of these initiation sources will definitely reduce the risk of explosion even though a dust cloud is formed. Dry atmosphere supports ignition of dust cloud. It is a wise move to keep humidity high in such areas prone to formation of dust cloud.

To summarize the safety measures.

Dust explosion or explosion due to the presence of volatile flammable fumes in the atmosphere can be prevented by

1) measures which prevent occurrence of ignition,
2) measures preventing source of ignition,
3) limiting damage using constructional design measures.

Usually a combination of all the three measures produces the best results, but (1) and (2) can be adequate in most situations.

Apart from fine powders being source of dust cloud during addition, layers of dust accumulated in process area and equipment is also a certain source for generating dust/air mixtures due to swirling of air currents. This can happen even when the fine RM is not being added. The pressure wave generated in a primary dust explosion kicks up the dust layers accumulated into another cloud and secondary explosions can occur which can be more violent. Therefore the importance of good housekeeping, removing dust accumulations in all areas of the plant, cannot be underestimated. Frequent cleaning and wetting with suitable inerting liquids can go a long way in preventing dust explosives from layered residues of additions.

Inert gases such as CO_2, water vapor, N_2, and noble gases can be used to blanket the area where dust clouds are formed to reduce the risk of explosion.

Monitoring of oxygen concentration and use of sensors are resorted to in closed systems of addition of fine powders to give warnings. Personal protection is through evacuation once warning is given.

Avoiding effective ignition sources such as welding, smoking, cutting of metals, mechanical/electrical sparking, and hot surfaces is a must. Not only energy content of the spark but its sources also have an influence on the probability of initiation.

Rotating steel parts have the capability to be sources of ignition and are governed by the hardness and relative circumferential speeds. Thus mixers/grinders/colloid mills come under equipment capable of producing mechanically generated sparks of sufficient high energy to be source of ignition to dust clouds or ignitable vapors.

8.2.3
Electrostatic Ignition

An electrostatic discharge or static as it is commonly known can set off dust cloud explosion if the energy released is greater than the minimum initiation energy of the mixture. The type of discharge determines the energy released in the spark. For example, brush discharge and spark discharge are different types of static one can encounter in industrial practices.

Principles of static prevention is grounded (no pun intended) in some fundamentals.

1) Ground all equipment, metal parts, metal drums, and interconnect buildings with conductors.
2) Ground persons working in hazardous areas – humidity, use of conductive flooring, wearing of conductive footwear, avoiding nylon-based clothing.
3) Reduce accumulated charge by using conductive material in drive belts and pipes.
4) Keep movement/flow velocities of liquids low. Avoid turbulence in atmosphere.
5) Use dust/explosion-proof motors and electrical fittings and remove accumulated dust regularly.

8.2.4
Lightning Protection

The conventional method is to have pointed metallic conductors on top of the building and connect them to earth. Clouds bearing electrically charged particles discharge their load into a conductor at its high point. The discharge is safely conducted outside the building into the earth. This method is not active but passive in its approach in the sense the upward leader propagates only after a long period of charge accumulation and reorganization. In proactive systems, the upward leader is created early by artificially generating through an ionization mechanism and the impact point of choice is established instead of waiting for a random strike. For the efficient performance of any type of lighting protection system, it is essential that a low impedance ground is provided to facilitate the dissipation of the lighting energy into the earth. Since soil conditions determine its conductance and this can vary from place to place, it is worthwhile to prepare artificial ground conditions as required and maintain it. Even here technology is available for long-term minimum or no maintenance systems. The use of such a system avoids frequent maintenance of the system which may be difficult in remote locations and to neglect such systems in good condition is equivalent to having no protection at all.

8.2.5
Runaway Reactions

Those chemical reactions, where the heat generated – exothermic – is greater than that dissipated to the surroundings, will tend to perpetuate exponentially and can be termed as *runaway reactions*. These are uncontrollable if left to themselves. The continuous increasing heat can lead to higher rates of the reaction generating more heat as rate of reaction increases with rise in temperature. Thus a situation arises where there is continuous increase in thermal energy which can lead to hazardous situation if left unattended. Although a specific reaction has been engineered to proceed under certain process conditions which is safe, operational deviations (accidental or deliberate) and control system failure can cause thermal explosions due to runaway reactions.

Factors primarily contributing to the incidence of runaway reaction are as follows:

1) side and unwanted reactions producing excessive thermal energy;
2) presence of substances in small quantities (unnoticed) catalyzing the reactions to go much beyond the intended rate of reaction;
3) inadequate process design for removal of heat generated in the chemical reaction;
4) failure of temperature controllers, flow meters, stirrers, coolant supply system, and solid addition feeders

In order to implement a process and operate it safely, thorough study of possible side reactions and their influence on the thermal energy balance needs to be done. Differential scanning calorimetry (DSC) and accelerated rate calorimetry (ARC) studies [7] need to be done to get this in-depth understanding of the reactions after which measures of control needed can be introduced. Choice between a batch, semicontinuous, and continuous process can be made looking at the potential safety risk in case of a runaway reaction.

Classical examples of dangerous runaway reactions are nitration reactions where end products are highly energetic materials such as NG and pentaerythritol tetranitrate (PETN) in unstable conditions. However these have been tamed to such an extent the production of such materials are not considered as beyond the limits of safety.

In case of manufacture of slurries and emulsions, there is no chemical reaction. It is merely a mixing operation as far as the actual production of explosives is concerned. The hazards of RM production processes need to be recognized and taken care of by the supplier of the RM.

The possible runaway reactions center around AN as this chemical is the major component. Extensive information is available in literature. A brief description of the reactions to be aware of when AN is subjected to heat is given earlier in Section 4.2.2.

The one most likely to become troublesome is the reaction of AN with sodium nitrite leading to unstable intermediate and further decomposition of the intermediate releasing heat.

This reaction can occur when oxidizer blend is being prepared. If the RM used for making the oxygen balance (OB) such as sodium nitrate, calcium nitrate, and sodium perchlorate contain excessive amounts of sodium or other nitrites, then a reaction could start between AN and nitrite. If the heat generated is not removed fast or the vessel is totally closed, then generated gas and heat could cause a hazard. Thus while making the OB, proper care needs to be taken regarding availability of cooling water and venting devices. The use of quenching as a tool to curb the runaway reaction is not practicable as the quantities involved are too large. A continuous check in temperature of the oxidizer blend as it is being made is a good check on unusual occurrences. A sudden rise in temperature should call for stoppage of whatever material is being added and cutting off the heating to the blend tank till the situation is under control regarding rise of temperature. Other runaway reactions possible are the reactions between AL and water which has also been discussed in detail earlier in Section 5.4.3.2. Of course the compatibility of any new RM being introduced in the explosive formulation with AN and other salts needs to be checked by using DSC/ARC [7] and information obtained on any possibility of runaway reaction and accordingly safeguarded [8].

The presence of water is to a great extent a deterrent for runaway reactions to occur and sustain. Since water content is at least 16–20% in most blends, there is plenty of water to absorb the heat generated and the tendency to form a runaway reaction is much reduced. Of greater concern is the situation where water has evaporated and dry mixes are formed. These when subjected to heat through friction can form precursors to a runaway reaction or explosion.

8.2.6
Venting as Means of Protection

Venting as a method to reduce the destructive force of an uncontrolled reaction is well recognized and established in the chemical industry and can be adopted in case of the explosive industry also, especially where runaway reactions and dust explosives are likely events. Venting does not prevent an explosion but reduces the damage due to pressure buildup in confined space by giving limited access to free atmosphere and the peak pressure is limited to a safe level capable of being withstood by the (chemical) equipment where the pressure has been generated such as a solution-making vessel, a reactor, a pump, mixers, and blenders.

The successful use of venting to limit explosion damage depends on the type of device used, its location, and its ability to perform as certified. The location should be such that the vented explosion has no serious secondary effect such as flying closure plates (metallic) or contents being thrown out to areas frequented by personnel.

Venting technology has advanced from the early times of spring-loaded relief valves, very large diameter bursting discs to explosion doors and panels, vacuum breakers, and one-piece rupture discs. These advances have made explosion-venting technology as an integral and vital part of the overall explosion protection methodology and have established it as a low-cost, viable alternative to explosion suppression and containment.

The basic principle of venting is release of pressure being generated in a closed system before it causes an explosion. Suppose an unusual event can generate a certain overpressure which could be more than the pressure the container has been designed to withstand under normal operating conditions, by fitting a vent device which opens at a lower pressure the pressure in the container can be limited to a pressure well below the design capability pressure and hence both the container and surroundings will suffer much less damage. For example, instead of disintegration into flying fragments capable of causing severe damage, bulging of the vessel could only be the damage suffered.

Explosion doors which can reclose after venting, air cushion explosion doors, vacuum breakers to prevent collapse of container, one piece rapture discs made of single domed metal membrane, encapsulated reed switch to activate opening of vent are all advances which have made explosion venting a reliable and predictable safety device available to plant designers.

8.2.7
Explosion Suppression Technology

As opposed to venting, explosion suppression technology [9] is effective in throttling the flame/thermal decomposition of an explosive occurrence as soon as it is initiated. The explosion suppression technology relies on early and instant detection of pressure, rate of increase of pressure, temperature, and visible flame as indicators to unload directly into the center of the occurrence liquid or solid suppressant suitable of overcoming the explosive event. If the suppression is successful, the explosion does not progress or proceed further and dies down. Since explosion is prevented in the very early stage itself no escape of detonation gases takes place and hence spread of toxic fumes is avoided. Early detection of an event characterized as potentially hazardous is achieved by microelectronic sensors which in turn activate instant release of suppressant. Multisensor detection systems can distinguish between explosions and process events. Use of high rate discharge suppressors are effective against dust explosions and gas explosion. Continuous development of highly sensitive devices for detection and improved suppressants in terms of specific effectiveness has contributed to the overall increase in use of such systems in the chemical industry. The explosive industry also, I believe, could be better off in terms of safety by adapting explosion suppression technology at appropriate locations.

Proper choice of suppressant is as important as its release. Monoammonium phosphate, sodium bicarbonate, and water are known to be effective in use.

Suppressants act best when they are having large surface area at the time of contact with exploding substance. Large surface area for solid suppressants means fine particle size and for liquid suppressants, fine droplet size in the form of a spray.

The theory of explosion suppression in closed vessels is based on deployment of an appropriate suppressant in more than adequate quantity into the combustion wave in order to reduce the temperature of the combustion zone and also to extinguish any flame in that vicinity. When this is effectively done, the combustion reaction cannot be sustained and normalcy can be restored. The suppression phenomenon is due to a combination of many other effects such as:

1) *Quenching*: Primarily, extraction of heat from the combustion zone by rapid energy (heat) transfer and by initiating endothermic reactions.
2) *Free radical scavenging*: Chemically active species in the suppressant molecule participate in chain-terminating reactions that compete with initial combustion reactions which propagate through a chain reaction.
3) *Dilution*: The suppressant used in excess dilutes the remaining explosives to such an extent that it is rendered noncombustible.

4) *Wetting*: Liquid suppressants render remaining explosives less flammable due to absorption and adsorption of nonreactive compound on the explosive.
5) *Disruption*: The combustion wave is physically disturbed by the impact of the suppressant and combustion wave suffers fragmentation and finds it difficult to propagate. The effects described above lower the combustion zone temperature, suppress the flame, and are very useful in the formulation of permissible emulsions of water gels.

8.3
Explosion Hazards in Equipment

While runaway reactions refer mainly to process, the equipment used to complete the process itself can be the location of hazard. The technology used in most of the processes for manufacture of AN-based civil explosives consists of mixing, pumping, conveying, and packing. Very few pieces of equipment are used for making AN/FO such as mixer, screw conveyor/feeder, and packer. The hazard analysis and history has show a low risk in the use of these equipment while producing AN/FO. The processes used in manufacture of slurries/water gels and emulsions are a bit more complex and involve high temperature and more sensitive ingredients. Process can call for high-energy input from high rpm mixer/colloid mill, conveying system in pipes using pumps and packing systems of high speed using pressure feed and closures.

Analysis of possible hazard locations and history of accidents in the last 10 years has shown that pumps are equipment with highest risk. Thus it is necessary to examine in detail the design of the pump and introduce effective in-built safety features [10, 11].

8.3.1
Hazards Associated with Pumping of Explosives

For pumping low-viscosity fluids such as oxidizer solutions or pumpable bulk products, double diaphragm pumps are found most suitable with high degree of safety. The material does not come into contact with moving parts. The diaphragm itself acts as a safety valve rupturing when overpressure is built up. Further in case of OB solution which is totally nonexplosive in nature and in case of bulk product, the line is open to atmosphere at one end and composition is also not very sensitive as the water content is high and the product is not viscous; hence the risk of explosion is minimal. It is claimed that diaphragm pumps can run dry indefinitely without damage. No shaft seals or gland packings are present and they are air-operated with no shearing action on the material being pumped. The pump stops immediately if discharge is clogged and there are no moving, sliding, or rotating parts. There are also no direct electrical connections to the pump reducing hazards due to electrical short circuiting.

Most problems have occurred when pumping highly viscous, low water content, cap-sensitive emulsions. Pumps identified as best suited for this purpose are positive displacement, progressively moving cavity pumps. These pumps take in the material in the cavities formed between stator and moving rotor, and during the running of the pump, the material is conveyed from one end (inlet) to the other end (outlet) and is pushed out continuously. Several situations can occur in this operation which can put extraordinary heat, friction, and compression on the product leading to explosions.

Different types of pumps tried out in water-based AN explosives [10] are

- gear pump
- lobe pump
- peristaltic pump
- piston pump
- diaphragm pump
- progressive cavity pump.

Irrespective of the type of pump used, basic causes for heat buildup and possible hazard risk are now established after detailed study of pump behavior using real explosive emulsions as pumped product. Such studies were carried out initially in Sweden, Norway, and Germany. Summarized below are hazardous situations which can happen in most water-based emulsion explosive pumping operations [11].

1) *Blocked inlet*: Closing the inlet valve by mistake or bridging of material even when the valve is open, lodging of lumps of solids and foreign bodies larger in size than the inlet valve opening can lead to blockage of inlet and stopping of flow of material to the pump. The pump if it continues to be operated with no flow will run dry soon and produces heat. Once the water in the emulsion is evaporated, the temperature increases rapidly and the dry residue can explode damaging the pump and surroundings.

2) *No feed from premix/matrix tank*: Here again after sometime the pump will run dry with possibility of explosion in the pump.

3) *Blocked outlet*: Outlet pipe is partly or fully blocked due to closure of valve by mistake or due to choking because of the presence of lumps, crystals, or solidified wax (due to drop in temperature of product suddenly). Pump produces heat due to pumping against increasing pressure due to outlet blockage. The heat produced is unable to escape and hence emulsion inside the pump gets heated continuously and provides the trigger for an explosion in the pump.

4) *Worn out rotor/stator*: Leakage due to excessive gap between rotor and stator due to wear and tear leads to leakage of product continuously leading to a situation where the explosive could be subjected to high friction. Air entrainment and cavitation leading to higher sensitivity of the product could create additional hazard.

5) *Mechanical breakdown*: Sudden breakage of shaft while pump is running will lead to dry running of the pump due to no output. Heat produced is absorbed by emulsion leading to a hazardous situation.

6) *Foreign objects inside pump*: Transported through product or introduced during open maintenance inadvertently or due to breakdown inside the pump (as indicated many times by unusual noise) can create extra friction, heat, and mechanical impact and the product may behave under these conditions more hazardously than usual.

7) *Pumping against deadhead*: Due to some event, the pump is kept running even when both inlet and outlet are closed or blocked but with some material inside. This is the worst scenario and creates maximum hazard and invariably results in an explosion at some point of time.

Early warning of the development of a hazardous situation goes a long way in preventing serious incidents while pumping. These are described next.

1) *Sensing switches*: Valves on inlet and outlet should be equipped with sensors. Only when the valves are fully open in the right sequence, signal is given for enabling pump to start and run. If no signal is received, pump will not start and also will shut down if valve is closed during running of the pump.

2) *Measurement of torque*: When pump is running normally, it will show a certain torque related to load on the pump motor. If outlet or inlet is blocked, torque will shoot up or drop down significantly. This indication can be used to stop the pump. Hence it is useful to install continuous torque measuring instrument and couple its output signal to pump motor operation.

3) *Measurement of pressure*: The pressure generated at normal operational parameters is taken as the standard value and is recorded by means of a sensitive pressure indicator. Significantly lower than standard pressure at the inlet side indicates lower feed and the pump shuts off on receipt of a signal. Pressure drop on the outlet side unless it is very sudden is not so serious and an audio indication is enough of a warning for the plant personnel to check and take action. If however the pressure drop is sudden, pump needs to be shut off. Response time should be fast so as not to allow pump to run dry.

4) *Measurement of temperature*: Dry running of pump gives heat and hence abnormal temperature results, and this can be noted visually or by a signal given to automatically shut off the pump. Measure of temperature difference of product between inlet and outlet is a good indication of any abnormality within the pump and its working.

5) Temperature should also be measured at the bearings and stator for a meaningful indication of heat rise due to mechanical problems.

6) *Measurement of flow*: Due to high viscosity and opacity of emulsion, measurement of viscosity on line is not very accurate and simple. Sophisticated flow meters can be, however, installed on line to give alarm signals to enable operator to check.

7) *Measurement of noise and vibration*: A good engineer will be able to easily tell whether the pump is working normally by listening to the noise it

makes and vibration it produces. Instrumentation in the form of audiometer and vibration measurement sensors are available for detection of improper working of the pump.

8) *Safety devices*: While the above described measures are indicators for unusual operating conditions and measures for stopping them before any serious damage occurs, one would not like a continuous process to get interrupted often and hence precautions need to be taken to see that many potential threats are prevented or reduced in the initial stages itself. Such devices are not very expensive and simple to install and operate but give a lot of protection against hazardous situations. These are as follows:

   a. *Filters on lines*: These prevent clogging of lines from foreign bodies and lumps of solids. One should note that filters can be effective only if they are frequently and thoroughly cleaned.

   b. *Level indicators*: These devices are quite easy to install and can be effective if their position is well above the empty level in tanks/hoppers containing materials that are being pumped. These give off alarm when level of material drops below safety level and can be hooked to stop operation of pump when that level is reached.

   c. *Water lubricated seals*: Removes heat generated due to unusual friction in bearings and can be rigged to stop the pump if there is no water being circulated in the seals.

8.3.2
Possible Hazards during Packing

Apart from hazards associated with pumping of explosives described above, another area of concern is the packing machines used for introducing the explosives into plastic or paper shells. Currently most of the slurries and emulsions are packed into plastic sheaths which are clipped at both ends. The machine used for packing small-diameter explosives, an offshoot of sausage packing machine, is a high-speed machine where the explosive material is continuously pushed into a plastic film and clipped. The speed of cartridging and the pumping rates are matched to regulate the quantity of explosive going into individual cartridge. The clipping mechanism consists of wire getting cut and preformed into clips, which are then further pressed over the gathered folds of the plastic film at the end of the cartridge. There is considerable pressure exerted in this closure operation in order to ensure that there is no leakage of material through the clipped ends.

The possibility of explosive material coming into contact with the closure plates during packing operation cannot be ruled out and is a definite matter of concern as the explosive material could then be subjected to frictional forces and pressures. The fact that most compositions of slurries and emulsions are insensitive to friction and impact forces gives some comfort in that the risk of explosion due to such stimuli is much less and hence the use of high-speed packing machines have become common. But here is not enough data, unlike in the case of pumps operation, to estimate the extent of hazard associated with the operation of high-speed packing machines. Although there have been some accidents reported during packing operations, it has not been possible to pinpoint the exact location of the initiation.

Apart from high-speed sausage style packing machines, there are machines which extrude stiffer compositions, not possible to pump, into preformed paper or plastic cartridges. A certain amount of pressure is required for this operation and normally since the line is open at the other end the force exerted on the material is minimal more so as the material is thixotropic and flows easily under pressure. The problem arises if the line is choked due to foreign object or lumps of hardened explosives. Such a situation is potentially dangerous as considerable pressure can be exerted on the material by the air-operated or hydraulic pistons pushing the material against a blocked space.

To prevent such a situation developing into a hazard, the piston operation needs to be stopped as soon as counter pressure is felt and the piston travel stops or slows down substantially. Sensors can be fitted to detect such an event happening. These sensors on feeling unusual resistance in the movement of the piston cut off the power to the piston.

8.4
Concluding Remarks

The ultimate objective of all safety measures is guided by the philosophy of giving priority to reduce direct and indirect exposure of human beings to hazard risks by reducing quantity of explosive handled and its exposure to hazardous stimuli. The purpose of all activities of an explosives facility is oriented toward

- supplying safe products to the end user
- maintaining at all times safe conditions of work
- ensuring a safe external environment.

The philosophy of safety management should be the implementation of total safety for the entire operation, although more emphasis obviously has to be on critical areas where higher levels of hazards exist. The total safety concept is best planned and introduced at the commencement of the project involving all aspects of producing an explosive. This gives an opportunity to scrutinize layout, safety distances, equipment (selection and layout), precautionary hazard fighting equipment and measures, their location, training, and selection of operating personnel. Above all commitment to the concept of safety as a way of life from the top management will go a long way in ensuring a safe operating organization.

References

1. Mahadevan, E.G. (2006). Safe, Safer. *Visfotak J. (India)*, **I** (1), 15–17.
2. Agrawal, J.P. (2010) *High Energy Materials*, Chapter 6, Wiley-VCH Verlag GmbH, Weinhiem, pp. 413–446.
3. Sreenivasan, N.S. (1983) *Management of Safety*, IDL Ltd, India, pp. 110–133.
4. Enoksson, B. (1981) Safety philosophy in the explosive industry. Proceedings of the 7th Safety Congress, Athens, Greece.
5. Rossel, S. (1988) Safety in Manufacture of Explosives, Internal Communication, Sweden.
6. Siewek, R. (1997). Dust explosion protection, *Perry's Chemical Handbook for Chemical Engineering*. 7th Edn, Mcgraw-Hill.
7. Mores, S. and Ottoway, M. (1998) International Seminar on Thermal Fire Explosion Hazards in Process Industries, IICT, India.
8. Khan, A.A. (1998) Runaway Reactions-Precautions and Safeguards, International Seminar on Hazards in Process Industries, IICT, India.
9. Siewek, R. and Moore, P.E. (1995) New development in explosion suppression. 8th International Symposium on Loss Prevention and safety Promotion in Process Industries, Antwerp, Belgium, Vol. I, pp. 539–550.
10. (1997) *Guide Lines for Pumping of Bulk Waterbased Explosives*, Institute of Makers of Explosive, Washington, DC.
11. Mahadevan, E.G. (2003) in *Theory and Practice of Energetic Materials* (eds P. Huang, S. Li, and Y. Wang), Science Press, Beijing/New York, pp. 873–881.

9
Economics of AN-Based Explosives

9.1
In Manufacture

It is inevitable that cost of setting up a manufacturing facility for ammonium nitrate (AN)-based explosives is compared with that of a nitroglycerine (NG) explosive manufacturing unit. At comparable capacities (tonnages of output) and product mix, the capital costs are substantially higher for a NG explosive facility due to the high expenditure involved in setting up a captive unit for manufacturing of NG, the most essential ingredient. By buying master mix from an outside source, the cost outlay can be reduced to some extent but not enough to be reaching the levels comparable with AN/FO, water gel, or emulsion plants. Of the three types of AN-based explosives, the least cost option is for the AN/FO unit but the product range would be limited to large-diameter and bulk explosives. There is no significant difference between the capital costs involved in setting up either a slurry/water gel plant and an emulsion unit of similar capacities and product range. The process involved for manufacture of both types of explosives needs storage tanks, oxidizer tanks, fuel blend tanks, mixer, packaging equipment, and auxiliary equipment such as pumps. Utilities can be generated in similar equipment costing about the same. Scope for reducing the capital expenditure is limited by the nature of the major raw materials being handled. Since hot AN solution is involved in an extensive way, corrosion-resistant material of construction needs to be used for most of the equipment all the way in the plant from AN melt storage to packing machines. Even the pipes conveying the AN solution or explosives have to be made out of stainless steel. Substantial cost reduction can happen only if new nonmetallic materials of construction are used for the process equipment. A strong possibility exists to start with in making hoppers, fuel blend tanks avoiding costly stainless steel. A trade-off between costs and life of the plant will arise at some point of reduction in the costs and the right decision needs to be taken at that time.

Running costs will vary from country to country depending on the cost of utilities and personnel. In industrially advanced countries, there are advantages by going in for additional costs on automation in order to save on manpower which could be very expensive. This may not be necessary in developing countries where cost of manpower could be low. A via media of semiautomatic plants could also be

Ammonium Nitrate Explosives for Civil Applications: Slurries, Emulsions and Ammonium Nitrate Fuel Oils,
First Edition. E.G. Mahadevan.
© 2013 Wiley-VCH Verlag GmbH & Co. KGaA. Published 2013 by Wiley-VCH Verlag GmbH & Co. KGaA.

an answer especially if exposure of people to explosives is to be limited even if it means more expenditure.

Minimum inventory of raw material, work in progress, and finished goods will keep the working capital required low and save on interest charge. Lean management techniques should be introduced to ensure greater efficiency and productivity so that running cost of production per ton of explosives produced is lowered.

Utilities costs are another source of expenditure amenable to cost control and reduction.

Steam generation in efficient boilers with the cheapest available fuel, reduction in steam distribution losses in steam lines by better insulation, design, and maintenance are all standard measures used in chemical plants which can also be used in the explosive units for the same purpose of controlling continuously rising costs. Absolute requirement of steam (quantity/pressure) can be reduced by lowering of processing temperatures without affecting quality and productivity of the explosive being produced. Apart from saving expenditure on steam generation, lowering of process temperature brings in a greater safety margin and reduces hazard risks in production operations.

Energy costs and its periodic escalation cause an increase in the utilities cost of production. Use of energy-saving devices for plant and surrounding areas' lighting, energy audit to prevent overcapacity/design of motors involved in process, reduction in friction load of moving parts due to wear and tear are some of the ways to lower energy costs and reduce per ton consumption of electricity.

Water can also be expensive in some locations especially if it needs elaborate pretreatment before use in process. Hence water conservation measures such as recirculation, utilizing steam condensate for process, and preventing evaporation losses will help in reducing the consumption of water which will reflect in lower utilities cost.

Since most formulations are close to each other in the percentage of main ingredients used, cost difference occurs only because of transportation cost of raw material and finished goods due to location of plant. Either the plant should be close to port/raw material source or it should be close to major customers. The raw material costs are always need to be looked at on a landed cost basis by the manufacturers whereas the customer looks at the landed cost basis of the explosive which he would like to buy.

Since a wide choice of raw materials can be used in slurries and emulsion explosives, continuous monitoring of the cost of the various raw materials available for use needs to be done in order to use the cheapest possible option. This applies equally to packaging materials which form a substantial part of the cost structure of an explosive especially in small-diameter packing (20%). Even a small reduction in costs per cartridge will multiply to a substantial amount since each ton of small-diameter explosive has 6000–8000 individual cartridges.

Reduction in manpower can form a substantial cost savings especially in countries with high wage structures. Simple devices can sometimes reduce manpower and increase productivity. But a significant planned low manpower requires design of a plant with automated controls operated from a central monitoring facility. Apart from the high capital costs for automation, availability of skilled suitably qualified manpower and essential spare parts on site is a must for the successful operation of an automatic plant. A continuous process plant lends itself as best suited for introduction of automatic online controls. This ensures both consistent quality and safety of personnel. The sensors, alarms, and instrumentation all need dust-proof atmosphere and operate more reliably if housed at constant temperature and humidity possible to attain in an air-conditioned atmosphere and not subjected to hot, varying atmosphere of a tropical climate. Air conditioning is a costly venture both in capital and running costs and is also heavily dependent on an uninterrupted power supply. In most developing countries, this may be difficult to ensure and semiautomatic plants with open civil structures are preferred and have been functioning efficiently and economically for the last two to three decades.

9.2
In Applications

Energy liberated when an explosive detonates is the main cause of work done on the surroundings. The success of an explosive in any field application lies not only in its ability to release energy fast and to its full potential (theoretical maximum) but also in the transfer of energy without loss to the surroundings. While the former is controlled by the explosive manufacturer through proper formulation and quality control of the end product, the latter is beyond the control of the manufacturer as it is in the hands of the end user to obtain the best performance out of the explosive. Having said that it is now well understood that a close working of the explosives manufacturer and end user is necessary over a long period to arrive at the optimum explosive and blast design for obtaining the most economical output, be it an opencast operation or an underground mining.

The mining operations can broadly be classified as

- open cast mining for coal,
- open cast mining for noncoal,
- underground mining for coal,
- underground mining for noncoal,
- excavation and tunneling,
- quarrying and well digging,
- trenching,
- underwater blasting,
- smooth blasting.

The requirements of the explosives, blast designs, and postblast results all vary for the different types of operations mentioned above. For example, whereas in mining for coal, medium-sized chunks are preferred as postblast physical requirement of the blasted material, in iron ore mining well-fragmented end product is desired. In both the cases boulders should not be present after a blast so as to avoid secondary blasting which will add to the total cost of recovery of ore.

The blast design employed by blasting engineers, whether in opencast and underground mining, are different and complex. Much work has been done to arrive at theoretically the best design followed by extensive field efforts of blasting to establish them in practice. There are many books and papers published on this subject and a few outstanding ones are mentioned in this chapter [1, 2].

It is not within the scope of this book to go into details of blasting science and technology, but it is necessary to understand and point out the possibilities of valuable explosive energy being lost in a blasting operation and how to avoid it.

Although the explosive cost may form only 10–15% of the total blasting operation, if the explosive energy paid for is not fully utilized during blasting then the cost of the operation is bound to go up and where a blast has totally failed the economic consequences can be quite severe. Quite frequently for such happenings the explosive manufacturer is penalized for supply of substandard or ineffective explosives. Hence it is in the interest of the explosive manufacturer not only to supply the right type of explosive but also ensure that it is in prime condition as regards its explosive properties. It is also best that the explosive manufacturer closely associates with the end user in evolving optimum blast design so that the explosive energy is fully utilized and delivers the expected results. Assuming that the correct explosive has been chosen for the blasting operation, factors which affect the end results (adversely) are as follows:

- Condition of explosive – velocity of detonation (VOD), age, air-gap sensitivity, strength
- Coupling – filling of borehole with explosive leaving no free annular space or voids in the column
- Priming – appropriate type and quantity of primer and position of priming
- Inert material – plastic film between explosive and interior walls of the borehole
- Stemming – quality and quantity
- Quantity of explosives loaded and its height in the borehole
- Blast design
- Water in the hole
- Explosive energy partition
- Explosive–rock interaction and confinement.

9.2.1
Condition of Explosive

It has been discussed earlier that explosive properties reduce with time depending on the formulation, method of manufacture, and storage conditions. Hence it is important to ensure that explosive in prime condition is used for blasting. This is done by using an explosive well within its shelf life. A quick check if in doubt can be done easily by measuring the VOD in the open using D'Autriche method and comparing it with that specified by the manufacturer. In case of bulk explosive, more reliance has to be placed on the density of the product and its composition as measuring VOD may be difficult. It can still be done by collecting the bulk explosive in a polythene tubing of the diameter being used for blasting and measuring its VOD on the surface.

9.2.2
Coupling and Priming

One of the major claimed advantages with soft and flowy explosive is its ability to spread and fill a borehole completely without leaving any annular gaps or voids in the column. This can be ensured by careful loading, especially if water is present in the borehole from the bottom upward. Contact with rock will ensure complete transfer of shock and bubble energy to the surrounding strata. No energy is dissipated into air or inert packing material. Whether coupling has been achieved in full can be checked by comparing the volume of explosive loaded with the volume of the borehole.

The effect of over/underpriming has been discussed earlier. Underpriming is more serious and affects the blast results severely, sometimes to the extent of total failure. This has to be avoided by use of primer in sufficient quantity, diameter, and detonation pressure (VOD). The primer itself should never fail but to have a fall back in case of failure, dual priming at different points in the column is resorted to. Care is taken to see that the second primer usually in the top of the column detonates only at the last stage of the propagation wave from the explosive has reached, otherwise collision of two waves moving in opposite directions can occur in the middle of the column and create spallation.

Primer position has been investigated extensively as it was thought to exert influence on the blast result. Most blast designs operate with maximum output if bottom priming is adopted and the detonation wave travels from the bottom of the hole to the collar. Multipoint initiation practiced, often because of practical difficulties, are not conducive to deliver the maximum energy of the explosive to the rock.

To set off the primer, detonating cord has been used in most blasting operations for reasons of ease of priming, low rate of failure, surface connectivity, and low cost. The most popular detonating cord consists of a load of 10 g/m pentaerythritol tetranitrate (PETN) and detonates at a velocity of 7000 m/s. The drawback of using the 10 g/m cord is that there is a possibility of the detonation wave disturbing the explosive charge as it moves down the column. It can also loosen the stemming material on top and even develop fissures in the rock. All these affect the blast performance adversely, especially the effect on the explosive charge. The reduced performance of the explosive is attributed to the pressure desensitization of the explosive by the shock wave from the detonating cord. This results in increase in density in the explosive and could lead to borderline sensitivity problems. Further the air bubbles in the body of the explosive could get destroyed and the number of hotspots could get reduced leading to a lower rate of reaction of the explosive decomposition and lower VOD than usual, which in turn will affect the energy release to the surroundings resulting in inefficient blast effect.

With the advent of the nonelectric system where a low VOD (1300 m/s) detonation wave in a tube conveys the initiation impulse to the primer, the problem of pressure desensitization of the column charge, multipoint initiation, and air blast noise pollution has been eliminated. The use of long-lead detonator for bottom priming has also been used with success but in deep holes of opencast mining it is somewhat expensive and failure due to electrical problems and hazards due to static have practically eliminated using this type of initiation in favor of detonating cord and nonelectric system. Only in underground mining it finds application. In order to avoid disadvantages of using a 10 g/m detonating cord, the manufacturers have also come out with 5 and 3.5 g/m charged detonating cords. These cords have less noise pollution but deliver the same detonation pressure since their VOD is almost around 6700 m/s. Their lower explosive charge mitigates the effect of pressure desensitization of the explosive loaded in the column. Their tensile strength is comparable to the standard 10 g/m cord. However the use of these cords containing lower explosive charge is still much less in practice than that of 10 g/m standard detonating cords.

9.2.3
Stemming and Confinement

The importance of stemming is many times forgotten while blasting. Analysis of blast event has, however, revealed the vital role played by stemming as extremely influential on the blasting results and should never be neglected. Stemming is to a great extent also responsible for momentarily confining the explosive inside the borehole to derive maximum energy transfer to the rocks before venting to the atmosphere to allow the expanded gases to escape. Poor stemming practice such as loose stemming, use of wrong stemming materials, and inadequate quantity (length of stemming) result in premature ejection of the stemming material which will affect significantly in lowering the quantum of energy available to do work on the rocks surrounding the explosives. The burden rock velocity which is a measure of the efficiency of the blast is reduced significantly by ejection of the stemming before the detonation has reached its maximum potential energy (borehole pressure). A blast with poor stemming also results in excessive fly rock and air noise.

Confinement in a blast is provided by the rock body surrounding the explosives in a borehole. An explosive with higher total energy is expected to function efficiently at a higher degree of confinement than an explosive of lower energy content. The former should be able to displace a greater burden. Overconfinement for all explosives will lead to no burden movement and results in a failed blast. Thus the amount of confinement has to be estimated and implemented by the blaster in the blast pattern adopted.

9.2.4
Explosives–Rock Interaction

The interaction of surrounding rock with the loaded explosive is an event where high pressure created by the expanding postdetonation gases, the temperature, and shock energy create physical movement and fragmentation of the rock. This interaction depends on the detonation properties of the explosives (VOD, volume of gas generated, explosion temperature, and energy density), physical and mechanical structural properties of the rock, and the confinement provided. This complex process of explosives–rock interaction is not easy to predict because of the many variables involved. The study of this phenomenon has engaged researchers in the field of blasting for many decades. The data generated has resulted in much understanding and has eliminated total guesswork and made it into a predictable event. Still there is not a perfect match between results predicted from theory and the various computerized programs used for calculations and field results mainly due to the variations in the rock strata which cannot be defined with accuracy [3]. Rock properties have a profound effect on blast performance and end results. Matching rock properties like compressive strength and sonic velocities with explosive properties has been continuously attempted and in many field blasts successful. There have been also less than desired results not easily explained. Overblasting and generation of fines have been encountered even while using standard and computer-aided designs. In some instances, underblasting and production of unacceptable number of boulders have resulted. Thus the blasting engineer is faced with the challenge of determining the correct match between explosives properties available and rock properties known to exist in the strata to be blasted and accordingly come up with a suitable design.

The prediction of blast results takes into consideration explosives properties, blast design, and the rock characteristics. The obligation of the explosives manufacturer rests in providing explosives with the desired physical and explosive properties and behavior in field conditions. Blast design is the responsibility of the blasting engineer and the rock characteristics evaluation that of a geologist. Thus a successful blasting event requires unified efforts of all three professionals and is a fascinating and sometime frustrating exercise in blast project management.

9.2.5
Explosives Energy Optimization in Blasting

Formulating an explosive with desirable properties does not ensure that blast results are satisfactory. The key to efficient blasting lies in proper energy distribution, in supply of the required energy level, and in the desired degree of energy confinement.

Energy level is decided according to the quantity and type of rock to be broken and accordingly the correct explosive chosen. The total energy level of an explosive or its energy content is realized in detonation in two components, as shock energy and as gas energy. The shock energy is dependent on the detonation pressure which in turn is derived from VOD of the explosive. The gas energy is obtained from the expanding gases related to the chemical decomposition of the ingredients in the explosive and the explosion temperature. Both these determine to what extent the gas is generated and how much it will expand and work. The partition of energy is generally in the ratio of 70 to 30 (gas energy to shock energy) for most AN-based explosives except for AN/FO where the ratio is 85 to 15. These ratios are from values obtained in underwater energy measurements.

The energy components need to be distributed and utilized in a correct manner to get the maximum benefit in a blast. The energy partition and distribution is decided based on the rock properties and geological composition of the strata being blasted. The energy is concentrated more at the bottom of the hole and in areas where a rock band of hard material exists. Concentration of energy at the bottom of the hole (about one-third of the column height) is achieved by using high-energy, high-density explosive. Care needs to be taken to see that while striving for high densities at the bottom, sensitivity to initiation is not compromised especially if the explosive is subjected to further pressures from the weight of the rest of the column charge acting on it.

A balanced energy utilization program yielding optimum blast results will consist of energy being allocated according to the quantum of work needed to be done, energy being properly confined to minimize loss of energy, and energy being of sufficient level to enable adequate movement of burden and fragmentation as desired.

9.3
Blast Design

Designing of a blast is purely the domain of the blasting engineer. After taking into consideration the properties of the explosive available and the geological structure of the rock to be excavated, the engineer has to decide on the right blast design which will accomplish the task of excavation at the lowest cost. In many cases getting the rock to move may be the primary objective. Whatever be the objective, the blast engineer decides on the blast design which takes into account the following:

- blast geometry
- burden and spacing
- stemming
- drilling pattern (number, arrangement, diameter, and depth)
- initiation system including delay
- quantity and type of explosive to be used
- air noise and environmental factors, ground vibration, and fly rock.

Blast is designed in such a way that postblast operation of loading, hauling, and crushing is done smoothly and efficiently. Today the explosives for civil blasting based on AN are fairly close in their characteristics. AN/FO has the lowest density and maximum gas volume. The emulsions show the highest VOD, slurries, and water gels are in between. If special requirements have to be fulfilled for a particular blast, the explosive will have to be made specifically for that event. Thus the blasting engineer has limited options from standardized explosives and keeping in view cost factor as important, a rule of thumb is adopted for selection of explosive.

Rock	Explosive
Hard	High strength
Medium	Normal strength
Weak	Low strength

In the earlier days, blasting parameters of burden/spacing/charge amount were arrived at after a number of blasts and standardized by trial and error. This was somewhat wasteful and depended much on the feel and intuition of the blasting engineer. Today the blasting operations are conducted on a very scientific basis [4]. Computer programs are available for arriving at the most desirable combination of explosives and blasting parameters by inputting data on explosive and rock properties of the strata to be blasted. The calculated values are endorsed by the accumulated field data and the computer programs are refined to take into considerations many other factors which may influence the blasting. Use of such computer programs to arrive at the most optimum blast design before actually conducting the blast eliminates a lot of unnecessary and wasteful efforts in conducting trial blasts and results in substantial savings. However since rock strata is never perfectly homogeneous and varies quite often, some deviations are bound to occur between the predictions of the computer program and the field result, but this gap is narrowed substantially by the refinements continuously being made to the software based on practical field data.

In order to really implement the most optimum blast design in the field and to observe its performance and classify it as successful, number of pre-, during-, and postblast measurements and observations are made. Some use sophisticated instrumentation and methods of calculation which can be practiced only by a research-oriented blast engineer. Simpler methods are also available to be used on a routine basis during regular blasting [5].

Preblast monitoring involves

- site characterization,
- ensuring right explosive of right quality,
- laser profiling of rock face, and
- environmental survey to ensure safe blasting.

Each of the above requires expertise if it is to be done correctly. For example, site characterization involves close examination and determination of the rock strata to be blasted. If it is understood that rock is not homogeneous and its properties will vary, still measuring of important rock properties such as density, compression strength, Young's modulus, hardness, Poisson's ratio, and sonic velocity using appropriate test methods and equipment covering the entire mine area needs to be done so as to have a correct idea of the site to be excavated. Geophysical logging by using exploratory drill holes initially for samples or using regular drill holes themselves for obtaining the samples and assessing the rock composition by geological analysis will supply good information about the site.

Feedback from operating drilling machines either through microprocessors or through experienced human operators will give data on the geological structure of the bench since drilling progress depends on rock strength (hardness/softness).

Laser profiling is done in relation to position of drill holes to identify zones of excessive burdens so that proper distribution of explosives can be done.

During the blast itself, monitoring [6] of certain parameters gives a feedback on what is happening in the blast itself.

1) measurement of VOD, borehole pressure in boreholes at various points;
2) high-speed photography of the blast;
3) measuring ground vibration near and far on critical structures needing protection; and
4) measuring gas penetration into rock.

High-speed photography and analysis of the data using 2D and 3D software gives information useful in blast diagnosis. Quantitative information can be obtained on

- firing times in individual drill holes,
- time for burden to move after detonation and stemming ejection time,
- velocity of burden movement,
- velocity of stemming ejection, and
- flight time of projectiles.

Quantitatively one can get information on

- misfires,
- firing sequence,
- fly rock, and
- oversize generation.

In-hole monitoring of VOD gives information on whether explosive has detonated throughout the column to its full potential or not, influence of primer (overdrive), and hydrostatic effect. Sophisticated electronic methods are available for measuring the *in situ* VOD in a borehole during a blast.

Detonation pressure generated during detonation can be measured using pressure gauges designed to withstand high pressures – these are known as *manganin gauges* and *piezoelectric gauges*. Monitoring air and ground vibration using detectors enables one to judge the hazards to the surroundings because of the blast and measures can be taken to keep it under control. From the explosives side, using less quantity explosives/primer, reducing surface blast effects by using detonation cord of lower strength or nonelectric systems, and using correct delay pattern (interval/sequence) can help in reducing excessive vibration.

Gas penetration measured through location of pressure transducers in the body of the rock defined by the last row of boreholes enables data to be generated, which gives relative assessment of gas energy generated by different explosives.

Postblast monitoring gives information on the end effect of a blast. These are

- fragmentation,
- profile of blasted material,
- throw of the rock, and
- muck pile looseness.

Fragmentation is monitored for size by several methods (direct and indirect):

- Visual analysis (actual counting of oversize)
- Digital photo analysis
- Loading equipment efficiency (shovel functioning)
- Crusher output (efficiency)
- Sieve analysis (Screening)
- Boulder count
- Quantity of secondary blasting.

Monitoring of digging/crushing equipment is by measuring torque/electricity consumption of motors used for the main function of the equipment.

Explosive performance as determined by its energy output measured in an underwater test is matched with the postblast results and adjustments made in quantity and type of explosive so that desired blast result is obtained. However it is to be noted that explosive is only one of the many important inputs influencing the end result.

Economics of blasting should be evaluated on a total cost basis but many times it is not. Cost of explosives per kilogram is many times the deciding factor in using a particular type of explosive. In such a decision, it will always be in favor of using AN/FO as it happens to be the cheapest available AN-based explosive. Not ignoring some of the advantageous explosive properties of AN/FO such as its high gas volume and low hazard sensitivity, still some blasting conditions especially those with water-filled boreholes require use of slurries/water gels and emulsions. The choice will depend on selecting an explosive system (explosive/primer/initiation system) which gives the lowest total cost for a blasting operation from start to finish consisting of the following steps:

- drilling
- primary/secondary blasting
- loading
- hauling
- crushing
- screening
- loading for dispatch.

All the operations are interlinked and controlling costs of individual operations in isolation may not produce the lowest total cost figures for the operation [7].

The drilling and blasting costs are quite significant in the overall costs. It would be possible to reduce these costs to some extent. By employing high-energy explosives and expanding the drilling pattern, the drilling costs would be reduced but explosives would cost more. There will be situations when these costs are very high because of hard rock but unless these operations are executed satisfactorily, whatever may be the cost, the subsequent operations will become very inefficient and their costs will go up affecting the overall project cost. Hence drilling and blasting always go together in achieving the correct fragmentation and throw so that subsequent steps involved function smoothly and efficiently [8].

An unusually low drilling/blasting cost may create problems in loading because of large boulder formation and increase the cost of loading to such an extent that the total cost becomes very high. Secondary blasting may have to be resorted to which again means additional expenditure and disruption of chain.

Operations to be avoided are

- increase in secondary blasting,
- low excavator efficiency,
- low dumper productivity,
- higher repairs to equipment, and
- blasting which leaves poor profile for the next blast.

Drilling accuracy and sticking to the established drilling pattern closely, using right energy distribution, use of delay sequences to create proper rock movement, and crushing by the gas and shock energy are basic principles of good blasting practice. Every blast can be a unique one due to changing strata and this has to be taken care during drilling/loading of explosives.

Data keeping accurately in all the steps mentioned earlier while blasting will enable one to observe improvements obtained by changing of some parameters.

9.4
Influence of Explosives in Underground Mining

While most of the factors discussed above are for large open cast mining, different practices can arise if the mineral mined is different (like iron ore or coal), but by and large the guidelines for good blasting practice remains the same as described earlier. However where the blasting is underground, different conditions will apply and accordingly blasting practices will change.

Underground mining, be it in coal or noncoal mine, although not involving huge volumes of explosives for blasting, provides challenging situations for the blasting engineer. Conditions usually prevalent for underground mining especially for tunneling for road and rail, hydroelectric power projects, mountain caverns for storage or other activities are different from those in open cast mining. In most cases, tunneling carried out below inhabited buildings will require careful planning to prevent damage from blasting to superstructures. But as in open cast mining excavation, underground mining except for coal involves different rock types and hence selection of appropriate explosives becomes important. Drill hole pattern, delay initiation, and sequence are vital to achieve good results which is usually measured as "advance" or "pull" in tunneling. In coal mining, the measure is yield in quantity with desirable fragmentation and minimum powder formation. Generation of postblast toxic fumes assumes critical importance in underground blasting even where adequate ventilation is provided. It is mandatory that postblast toxic fumes quantity is below a certain level immediately after blasting and the fumes are to be cleared within a stipulated time so that the face is approachable. Avoiding overbreak damage to roof (pillars) and the gallery due to excessive air/ground vibration and unutilized gas energy has to be avoided in order to reduce the cost of operation or compromise safety.

Modern techniques of tunnel blasting and excavation are executed such that entire excavation is done in a single operation known as *full face tunneling*. Large cross-sectional areas are involved in a single round of blasting. However underground coal mines, because of fragile roof, depth of operation, and loose strata, still operate with small face sizes. Moreover the drill hole diameter, type, and quantity of explosive per shot are predetermined and limited according to the gassiness of the coal seam being blasted. These constraints are much less in existence in noncoal underground mining and tunneling in projects where a continuous push for the use of larger diameter and deeper boreholes and use of greater energy per hole to obtain faster progress exists.

Tunnel blasting is technically challenging [9] and difficult conditions exist for the explosive to perform since only a single free face is initially available for the blast to progress. In comparison bench blasting can provide two to three free faces. Tunnels initially have only a single free face but additional free faces are created by delay blasting of drill holes in various patterns. The specific consumption of explosive is rather high (kilogram explosive per cubic meter of drilled hole) and is three times more than in bench blasting. Blasting result in tunnel is considered satisfactory if the advance is 90–95% of the drill hole depth. Here the best results are obtained with the explosive of high energy density which can perform at its maximum potential even in very small diameters.

Several factors responsible for good working results are similar to that seen in open cast mining such as rock properties, explosive properties, speed of reaction in the decomposition reaction, and position of initiation.

The economics of tunnel blasting is heavily dependent on charge factor (consumption of explosive/cubic meter), and drilling costs needed for securing a particular rate of advance. The role of explosive is to provide adequate breakage safely both in terms of peripheral structures and toxic gases and in case of gassy coal seams safety against ignition of methane/air and coal dust/air mixtures. Amount of energy in the explosive and its rate of release are critical factors for selecting an explosive for underground operation. Lower VOD explosives with more gas energy are preferred in underground coal mines in order to obtain manageable chunks of coal in the blast rather than fines. In noncoal mining and tunneling strength, oxygen balance, high density, high VOD of the explosive, and ability to perform in very small diameters (sometimes even 7/8 in.) are sought after in the explosive. No doubt it is easier to meet these criteria in larger diameters and hence the continuous push for usage of larger diameter boreholes and higher density explosives for blasting in tunnels.

Theoretical calculations of strength of explosive using BKW code [10], TIGER code [11], and Kuz–Ram model [12] are useful to know the ideality of the explosive being offered by the manufacturer. The explosive should not exhibit

- partial or low-order detonation or deflagration,
- sympathetic premature detonation in adjacent holes,
- channel effect, and
- dead pressing due to pressure wave from delay blasting.

If such an event happens in underground blasting, restoration of normalcy takes time and is costly because of restrictive working conditions. The presence of undetonated explosive could create safety hazard also but in this, the advent of AN-based explosive as substitute for NG explosives has reduced the risk considerably. Another difficulty in tunneling is that drill holes can be drilled only perpendicular to the face with no free face available. The explosive charge required will be higher than usual and the release of energy needs to be faster to prevent a blown out shot. There are many different ways of avoiding this type of predicament mainly based on drilling pattern, using delay sequence which creates free face as the firing progresses in the round. An explosive which is not dead pressed by passage of a detonation wave from a nearby hole or prematurely explodes is mandatory. Techniques which control overbreak at the perimeter require explosive energy to be distributed over a larger area by decoupling in the borehole. Low-density charge can also be employed if channel effect is feared.

Considering the required explosive properties for speedy advance in tunnel excavations, even though NG explosives have the highest density due to their tendency for sympathetic detonation, toxicity of postblast fumes, and low handling sensitivity, their usage has drastically reduced in the last 10 years. The NG explosives have been replaced to a great extent by water gels and emulsions. These show low toxic gas generation are safer to handle and have higher VOD than NG explosives. The density is the only property where the AN explosives fall short (1.25 g/cc against 1.45 g/cc) when compared to NG explosives. Continuous efforts are being made by R&D scientists to overcome this disadvantage at acceptable cost.

One of the factors affecting transfer of energy from explosive to inner walls of the borehole is the presence of packaging material (polythene) used to form cartridges. Following the tradition of NG explosive and to satisfy the end user's desire to have a substitute product similar to the NG explosive, the explosives manufacturers attempted and succeeded in providing AN-based water gels and emulsions very close in consistency (touch) to NG explosives. The packaging material may or may not participate in the detonation reaction and hence no effective contribution to energy can be expected from them. On the other hand they come in between the explosive and the walls of the borehole and delay the transfer of energy. The newly introduced pumpable explosives for small-diameter blast holes (gels and emulsions) avoid this problem and provide intimate coupling and efficient energy transfer. Their physical properties are in terms of viscosity such that even when loaded into horizontal or downwardly inclined holes they do not come out but continue to be fully inside. Such products are also sensitive to detonator initiation even in small diameters. They, however, need specially designed mixing and loading equipment. Sweden (Nitro Nobel) has taken the initiative and has put into practice such systems of pumpable products for small-diameter blasting in underground mines. The advantages are obvious – cost of packaging which is substantial in small-diameter explosives is eliminated. Even after considering the capital expenditure for the loading equipment, the overall cost advantage is definitely in favor of pumpable systems. It is surprising that the progress in converting packaged product into pumpable product for small diameters has been very slow. It should have been much faster considering the already available accumulated knowledge from bulk explosive systems. Carrying substantial inventory with its carrying costs, pilferage would get eliminated by using a pumpable system. It seems rather difficult to understand that considerable expenditure is incurred in equipment and materials to pack a product which later on serves no useful purpose in blasting.

References

1. Langefors, U. and Kihlstrom, B. (1963) *The Modern Technique of Rock Blasting*, John Wiley & Sons, Inc., New York.
2. Persson, P.A., Holmberg, R., and Lee, J. (1993) *Rock Blasting and Explosives Engineering*, CRC Press, Boca Raton, FL.
3. Mohanty, B. (2009) Intra hole and in-terhole effects in typical blast designs. Proceedings of 9th Symposium of Rock fragmentation, Granada, Spain.
4. Sarathy, M.O. (2000) Optimum blasting in surface mines, major issues. *International Seminar on Blasting Objectives and Risk Management*, IDL Industries, Hyderabad, India.
5. Bekkers, G. (2009) A practical approach to fragmentation optimization at Kidd Mine. Proceedings of 9th Symposium of Rock fragmentation, Granada, Spain.
6. Sarathy, M.O. (1994)Recent advances in surface blast monitoring and evaluation techniques-a review. *Indian Min. Eng. J. (India)*, **3**.
7. Grouhel, P.H.J. and Kleine, T. (1992) Designer-a spread sheet for determining the most cost effective explosives for an open cut blast. 3rd Large Open Pit Conference, Mackay, Australia.
8. Kendricks, C. and Malcolm, S. (1990) Integrated drill and shovel performance monitoring towards blast optimization. Proceedings of 16th SEE Conference, Orlando, USA.
9. Johansen, J. (1998) *Modern Trends in Tunnelling and Blast Design*, IDL Industries, Hyderabad, India.
10. Mader, C.L. (1967) A Code for Computing Detonation Properties of Explosives. Report LA 3704, Los Alamos, NM.
11. Copperthwaite, M. and Zwister, W.H. (1973) *TIGER Computer Program, SRI Publication No. 2106*, Stanford Research Institute.
12. Cunningham, C.V.B. (1982) The Kuz-Ram model for prediction of fragmentation from blasting. 1st International Symposium Rock Fragmentation, Lulea, Sweden, pp. 439–453.

10
Current Status and Concluding Remarks

There is no doubt that in the last decade or so, ammonium nitrate (AN)-based explosives especially water gels/emulsions have established themselves as mainstay in the field of civil applications where explosives are needed and used.

This status has been achieved mainly due to the simple raw materials required and the ease of production in large quantities in a safe manner. The explosives manufacturing and sale itself have undergone a remarkable change. Whether this change has been brought about by the availability of AN-based explosive or vice versa, it is a matter of academic interest. In practice the changes which are most noticeable are as follows:

1) Shifting of manufacture from a very large production unit in a single location to smaller facilities closer to the point of use. For example, as late as 1974 India had only three manufacturing factories producing explosives but in a matter of 20 years 30 more production plants have come into existence located all over the country. A similar situation has also taken place in China and even in the USA.

2) The importance of long shelf life (1–2 years) insisted upon in the earlier times has given away to very short period of 4–6 months. This has happened due to the change in the distribution system with the advent of bulk delivery systems and also very fast movement of packaged explosives. Customers carry very little inventory and the explosive is delivered just in time or made on the site itself. This has reduced the costs involved in storage and also enhanced the safety by reducing handling and pilferage. Further due to the wide geographical distribution of manufacturing plants, the explosives can reach an end user within 4–5 days and used up immediately thereafter. This ensures that the explosive's age when used is well within its stipulated shelf life and the explosive's performance is optimal. A shelf life of maximum 6 months seems to be adequate for most situations and accordingly the formulation and process can be tailored. It is only in case of export involving longer transit time due to shipping and storage at various stages that the explosive needs to have a longer shelf life of 1 year. This is somewhat of a paradox in that while R&D was all along busy in discovering formulations and processes to enhance the shelf life of the product (AN explosives) to be the same as nitroglycerine (NG) explosives,

Ammonium Nitrate Explosives for Civil Applications: Slurries, Emulsions and Ammonium Nitrate Fuel Oils,
First Edition. E.G. Mahadevan.
© 2013 Wiley-VCH Verlag GmbH & Co. KGaA. Published 2013 by Wiley-VCH Verlag GmbH & Co. KGaA.

the changed manufacturing and distribution systems allowed a much shorter shelf life for the product.

3) Change from the critical status given to explosives in the earlier days during blasting to a "commodity" status now where the explosive is treated only as an input in the chain of operation. Proliferation of explosives manufacturing units, some dedicated to a particular end user, has resulted in this situation. While the end user is benefitting from an assured and economical supply of the explosive for all times to come, there is a definite possibility that the full potential of the explosives which can be made available is not exploited.

4) Improvements in output from blasting activities of all types through implementation and execution of scientifically developed blast designs have reduced the dependence only on the energy-based performance of the explosive.

The question now is what direction the explosive industry needs to take in order to progress (survival is not the issue) and attain a preeminent position with the end user which status it was rightfully occupying in the early years. In particular, the areas of R&D have to be chosen carefully to attain this objective.

The explosive industry will need to concentrate on special applications and unique mining situations rather than routine blasting. With the growing population and habitation, the proximity of these to the mining operations will give rise to environmental problems which will need to be solved by special techniques using tailor-made explosives together with innovative blast designs. Thus it is again the close working together of the explosives chemist and the blasting engineer that will lead to solutions which are effective and acceptable.

The industry for its own sake can look at further enhancing the safety margins in all its operations. The safety statistics collected over the last decade on accidents involving the AN-based explosives production and movement is not particularly satisfactory and not commensurate with the safety margin expected from their fundamentally insensitive properties.

The scarcity of land surface eventually will lead to underwater mining and explosives will have to be developed specifically for this application. Although explosives are available for underwater applications such as deepening of harbors and channels, deep mining explosives for use in depths of more than 30 m underwater have still to be designed and tested. This may call for a totally new design concept from the existing products.

As regards underground mining especially in coal, explosives use and consumption will come down as mining practices will employ long wall methods using machines for coal cutting and dislodging coal without the hazards of continuous blasting using explosives. Even in opencast mines where strata is loose and soft like lignite, machine excavators are in use since long and their usage would increase. The major reason why these machines are not used more is because of the huge capital outlay required for such machines, the payback being risky due to the cost of mining, and fluctuating ore prices.

Therefore while the situation for the explosive industry is stable, any spurt in growth can come only from a revolutionary product discovery or opening up of new areas for use of AN-based explosives in large volumes such as military ordnance. Specialized applications such as metal cladding, metal forming, and seismic prospecting while technically challenging consume very much less quantities of explosive as compared to conventional mining and excavation activities. Already a decade has passed since the emulsion explosives came into regular commercial use; hence time is ripe for another breakthrough in the field of explosives for a new and revolutionary product. Perhaps microemulsions and nanotechnology could be the areas to spawn such a new product. Even if such an event happens, it is more likely to be of a technical novelty rather than a commercially endorsed breakthrough.

Appendix A

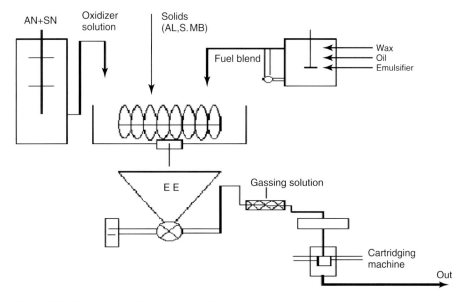

Figure A.1 Batch process for emulsion explosives manufacture.

Ammonium Nitrate Explosives for Civil Applications: Slurries, Emulsions and Ammonium Nitrate Fuel Oils,
First Edition. E.G. Mahadevan.
© 2013 Wiley-VCH Verlag GmbH & Co. KGaA. Published 2013 by Wiley-VCH Verlag GmbH & Co. KGaA.

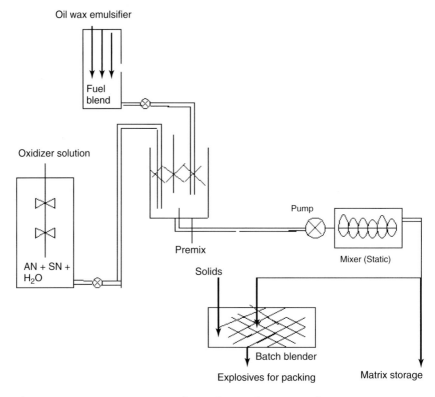

Oil wax emulsifier

Fuel blend

Oxidizer solution

AN + SN + H$_2$O

Premix

Pump

Mixer (Static)

Solids

Batch blender

Explosives for packing

Matrix storage

Figure A.2 Semicontinuous process for emulsion explosives manufacture.

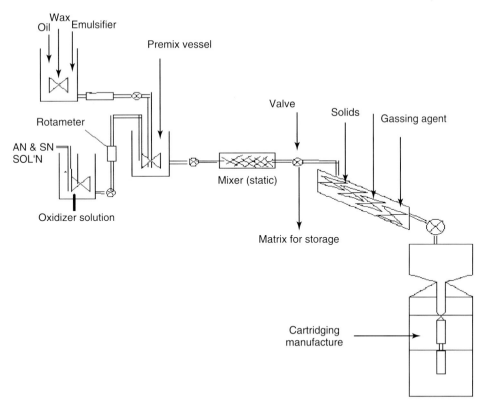

Figure A.3 Fully continuous process for emulsion explosives manufacture.

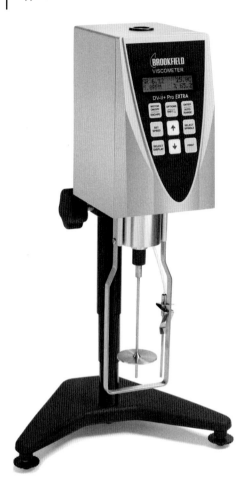

Figure A.4 The Brookfield viscometer.

Appendix B: Guidelines for Investigation of an Accident

B.1
Introduction

In spite of all the efforts in introducing appropriate safety measures and training people for a safe operation, accidents do happen sometimes with severe loss of lives and property. An important aspect of an accident which should not be neglected is the postaccident investigation. A properly conducted investigation is capable of eliciting a number of important facts which could lead to the main objective of the investigation namely identifying the root cause of the incident and how to prevent it from happening again and in the event there is a repetition, how to reduce damage to people and property.

The difficulty of finding evidence during investigation of an accident is increased with the severity of the explosion. In such cases where fatalities have occurred, it relates invariably to the people who could have given the best eye witness account of the happening not being available for giving an accurate account of the incident. In spite of these difficulties, experience built up over decades of accident investigation allows one to establish certain basic guidelines and procedures to be followed for obtaining the best results.

Assuming that an explosion has taken place in a manufacturing facility or a magazine usually located in an isolated area away from residential areas, the following sequence of events have to be executed:

1) Conveying briefly/accurately information of the accident to the concerned authorities such as explosive department, firefighting unit, owners of the unit, local police, local disaster management organization if any, factories inspectorate, and kith and kin of the people deceased or hurt in the accident. Avoiding communicating any information directly to the media at this juncture is a good practice in order to prevent wild speculation as to the incident in terms of its cause or extent of damage, and so on.

2) First impression: First impression (visual) of the accident site is to be obtained from personnel who are involved in the operation immediately after the incident as they will be close at hand.
First impression is necessary to know the effect of explosion immediately after the event. The following needs to be recorded:

Ammonium Nitrate Explosives for Civil Applications: Slurries, Emulsions and Ammonium Nitrate Fuel Oils,
First Edition. E.G. Mahadevan.
© 2013 Wiley-VCH Verlag GmbH & Co. KGaA. Published 2013 by Wiley-VCH Verlag GmbH & Co. KGaA.

a. Local weather conditions – temperature, humidity, wind velocity and direction, raining, or dry
b. Particular attention needs to be observed/recorded about the presence of lightning/thunderstorm and its proximity to the site of the accident
c. To observe/record color (white, gray, or black) of the smoke/fumes from the area of the explosion
d. To observe/record positions of major pieces of equipment
e. To observe/record visually physical damage to buildings and vehicles
f. To observe/record the physical characteristics and location of craters found in the area of the incident

Photography and visual observation needs to be resorted to record from a point closest to the site but considered safe from any residual explosive being present. The accident area should be cordoned off and no person should be allowed to enter and vitiate the area by movement. However this is easier said than done. In case of accidents involving fatalities to unearth buried humans and possibly to rescue them alive, there is a rush to the site of personnel concerned with such operations and invariably the site gets disturbed by search including the use of diggers and dozers. This is why the first impression of the scene however gross is a must and must be done immediately after the explosion.

To completely account for human damage, the plant in charge has to prepare (i) a list of personnel present inside the operational area – names and rank and job title, (ii) take head count with identification from supervisors to reveal missing/dead, and (iii) classify numbers under injured/dead/missing and update the list on a daily basis.

1) The licenses held, type of explosives being manufactured, quantity and description of Raw Materials (RMs) held, work in progress (quantity and type), and finished goods on site needs to be put down in the report.
2) Site plans (preaccident) needs to be available for the investigator to compare with the postaccident site to identify and record the damage occurred to civil buildings.
3) Similarly equipment layout with process flow sheet describing the operation before the incident needs to be available for examination.

A brief description of process/composition and ongoing operations at the time of the accident should also be drawn up.

B.2
Detailed Inspection

If possible before search, otherwise immediately after, emphasis on

- crater formation (location, size, width depth circumference, and shape);
- damage to civil structures estimated in terms of low or severe based on rubble size;

- leaning direction of buildings, equipment, towers, and compound walls;
- position of mobile vehicles (shift from original), extent of damage and location of damage on the vehicle, and any crater below the vehicles;
- details on the type of construction of each civil building and damage type and severity;
- quantity of explosive/RM/semifinished explosive (matrix) best estimate from records or from knowledge;
- position and quantity of RM in the storage building, especially those classified as fire risk such as FO, sulfur, AL, organic sensitizer, and AN;
- position and number (estimated) of personnel inside the danger area;
- location of equipment, pieces to be marked on a drawing. If identified, the distance from its original position, for example, lid of a vessel, shaft of a pump, part of a silo. Such equipment after marking to be kept safely for reconstruction of accident;
- damage to civil/equipment of neighboring industries/buildings, direction, and severity of damage to be noted.

B.3
Interviewing and Questioning

The best impression is obtained when done immediately after the incident provided the person is not in a state of shock and has clear thinking ability. Otherwise one should interview them later when they have recovered.

Eyewitnesses both in the plant and outside are helpful in piecing together the sequence of the incident but consistency in their observations is not always obtained. These observations should concern the severity of the explosion (volume of sound and number), presence/growth speed of fire and whether fire preceded explosion or vice versa, fumes generated (color, smell, and volume).

Enquiry of personnel as to whether any unusual activity such as welding, open fire operations such as burning/cooking of food was observed, sometimes elicits interesting possibilities.

- Sample of air to be gathered after the explosion and analyzed.
- Log book entry to be checked for any unusual phenomenon recorded such as abnormal rise in temperature, unusual equipment noise, and so on.

B.4
Collection of Samples

- Air sample as soon as possible (immediately) after explosion for analysis.
- Collection of any matrix/explosive/explodable material/sensitive RM which may have been strewn around after the explosion. Sample collected if sufficient in quantity to be tested in 3 in. diameter for cap sensitivity and velocity of detonation (VOD).

- Chemical analysis of sample to be verified against specification.
- Residual explosive product after collection of sample should be mopped up and the waste destroyed by open burning in a pit faraway from habitation. Initiating products such as detonating cord to be carefully gathered and destroyed by coexplosion in a deep pit surrounded by rubber mats to cushion the effect of explosion.

B.5
Examination of Witnesses

Cogent scenario to be put together after interrogation.

B.6
Examination of Dead/Injured

Postmortem report and report from examining doctor needs to be obtained and analyzed. Nature of damage, its location on the body, severity, and type can reveal the effect of pressure wave, its intensity, and direction. Burn injuries reflect the presence of fire. After obtaining all the information as above, the investigator has to analyze the data collected to come to a reasonable conclusion. The first objective is always to identify the initiation source and to pinpoint the location of the initiation event. Generally the initiation event is of no great sound or damage. There is hardly any crater formation. The only clue could be destruction/significant damage of equipment. It could be supported by rise of temperature/pressure. After initiation, the detonation wave will travel in all directions as long as there is explodable material present which will support and propagate the initial detonation wave continuously. This is usually done by explosives present in the pipes in a continuous column. The severity of the accident is caused by the reaching of the detonation wave to a stock of explosive held in storage. Depending on the quantity of explosive which is set off by initiation from the propagating/moving shock wave, the severity of damage is seen. The more the quantity of explosive setoff, greater is the damage. Confinement adds a degree of enhancement to the severity of damage and the explosive shock wave splinters the material in which the explosive is confined. Steel tubes transporting the explosive especially in hot condition act as transmission conduits. If the explosive setoff is in a large quantity, its detonation can set off sympathetic detonations of explosives/explodable substances in proximity. Well-known safety distance calculations are available to estimate the distances of separation needed to prevent sympathetic detonation. Many times even when the receiving explosive is outside the safety distance it can deflagrate, burn, and in confinement due to the action of escalating heat, it can explosively decompose.

Thus secondary explosions have been recorded after substantial time lag.

The formation of a crater of considerable size is seen in all detonation where a substantial amount of explosive is involved in the primary/secondary explosive. The size of the crater is more if the explosive is in good contact with the floor. In a bulk delivery vehicle or a silo where the explosive is 4–5 ft above ground, the crater formed is much less than that if the explosive was in close contact with the ground.

The direction of the shock wave is uniformly circular for a spherical charge. In case of mass explosion, the direction of the detonation wave is channeled where there is a measure of weakness like a light roof or a blast wall. On receipt of a shock wave, a receptor/building caves in or bulges out depending on the direction of the shock wave. Some indication as to the direction of the shock wave can be obtained by looking at the direction of the fall of the debris. Firm conclusion cannot be drawn only on this basis as to the source of explosion since a suck back occurs after the initial pressure wave has gone through. This suck back can show damage in the direction opposite to that of the traveling shock wave. More accurate information is obtained from the distance/direction the equipment has been dispersed.

Thus the study of the debris, crater, equipment damage, and injuries should give enough data to pinpoint source and location of explosion with reasonable certainty. However it is to be kept in mind that each incident is unique and generalizations can mislead.

Index

a

accelerated hot storage (ageing test) 27–28
air bubble sensitization 103
air entrapment and occlusion while
   emulsification, by mechanical agitation
   125
air/gas/synthetic bubble sensitizers 108–111
aluminum in water gels and slurry explosives
   74
– aluminum-based water gel explosives, tests
   for AL powder for use in 80–84
– aluminum water reaction 78–80
– atomized powders 74–77
– flake powders 77–78
amphipathy 132
anionic surfactants 133
apparent viscosity 97
aquarium test 23
Audibert Delmas test method 100–101

b

ball drop method 140, 141
ballistic mortar test 18–19
BAM Koenen test 101
beeswax/borax combination 129
BET analysis 78, 82
black powder 5
blast design 186–190, 196
blasting agents 15
booster-sensitive explosives 1
Bruceton up and down method 87
bubble energy 8
bulk delivery product 61
bulk delivery systems 40–41
bulk delivery vehicle 41
bulk density 42–43
bulk emulsions 149–151
butadiene-styrene 124

c

caking phenomenon 36
calcium nitrate (CN) 62–63, 138
cap-sensitivity explosives 1, 15, 16, 18, 19, 21,
   22, 26, 27
cartridged explosive 3
cephalins 129
Cerchar's method 101
characteristics, of ANFO 45
– density/strength 45
– detonation propagation mechanism 49–50
– energy content 46–47
– fuel influence 50
– increasing of energy and fume
   characteristics 52–55
– moisture /wet boreholes/water-resistant
   AN/FO 50–52
– strength of explosive 46
– velocity of detonation and effective priming
   47–49
– water-resistant AN/FO 52
chemical gassing 125–126
chronology 15–16
chub style packing 69
classification, of explosives 1
– initiation sensitivity 1
– physical form 3
– size 1–2
– usage 2
cold temperature storage test 28
commercial explosives, design of
– oxygen balance, importance of 16–17
– physical, performance, and safety
   requirements 17
continuity of detonation (COD) test 22–23,
   87, 136
cooling, of emulsions 136
crater test 26

Ammonium Nitrate Explosives for Civil Applications: Slurries, Emulsions and Ammonium Nitrate Fuel Oils,
First Edition. E.G. Mahadevan.
© 2013 Wiley-VCH Verlag GmbH & Co. KGaA. Published 2013 by Wiley-VCH Verlag GmbH & Co. KGaA.

critical diameter   9
critical scaled depth   26
cross-linking   95–96
crushing strength   43–44
crystal habit modifiers   127–128
crystallization point   140
crystallographic studies   24
cup density   45, 85
cylinder (crushing test) test   24

d

D'Autriche method   21, 22
decomposition chemistry   32–33
deflagration   5
– mass limit   101
– tests   100–101
detonating fuse   21
detonation   9
– pressure and velocity   9–11
– propagation mechanism   49–50
detonics   10
dielectric measurements   142
Differential Scanning Calorimetry
   thermogram   57
dilution   171
doped emulsions   151
doping   106
double-cone blenders   40
double pipe test   23–24
drilling test   104
dust explosions   91

e

economics, of explosives   179
– in applications   181–182
– – coupling and priming   183–184
– – energy optimization in blasting   185–186
– – explosive condition   182–183
– – explosives–rock interaction   185
– – stemming and confinement   184
– blast design   186–190
– explosive influence in underground mining
   190–193
– in manufacture   179–181
Ecospheres   127
electrostatic ignition   167–168
emulsion explosives   113
– bulk emulsions   149–151
– composition and theory of   117–118
– concept of   113–114
– emulsion matrix properties   142–145
– – channel effect   144–145
– general composition of   114–115
– GPSD emulsion explosives   147–149

– HANFO   151–152
– manufacture
– – air entrapment and occlusion while
   emulsification by mechanical agitation
   125
– – chemical gassing   125–126
– – crystal habit modifiers   127–128
– – emulsion chemistry and understanding
   emulsifiers   129–133
– – emulsion promoters   128
– – emulsion stabilizers   128–129
– – fuel blends   123–124
– – HLB and use in emulsification   133–138
– – hollow particles   126–127
– – manufacturing process   118–122
– – polymer systems in emulsion explosives
   138–139
– – product types   118
– packaged-large-diameter emulsion
   explosives   153–154
– permissible emulsions   145–147
– quality checks
– – process audit   140–141
– – raw materials   139–140
– – special tests   141–142
– structure and rheology of   115–117
energy optimization in blasting, explosives
   185–186
enthalpy   11
ethyl vinyl acetate (EVA)   124
exfoliation   145
Expancel   127
explosives, of ammonium nitrate fuel oil   39
– characteristics
– – density/strength   45
– – detonation propagation mechanism
   49–50
– – energy content   46–47
– – fuel influence   50
– – increasing of energy and fume
   characteristics   52–55
– – moisture /wet boreholes/water-resistant
   AN/FO   50–52
– – strength of explosive   46
– – velocity of detonation and effective
   priming   47–49
– – water-resistant AN/FO   52
– manufacture
– – bulk delivery systems   40–41
– – continuous process   40
– – mixing process and equipment   39–40
– properties
– – oil absorbency and porosity/bulk
   density/crushing strength   41–44

– – physical   41
– – temperature cycling effect resistance   44
– quality checks   56–58
– safety considerations   55
explosive science   5
– detonation   8
– – ideal/nonideal detonation/critical
   diameter/ideal diameter   9
– – pressure and velocity   9–11
– high explosives   5
– initiation and detonation
– – mechanism   6–7
– low explosives   5
– propagation   7–8
– reaction chemistry
– – heat of reaction   11–12
– – hierarchy rules   12
– – oxygen balance and fuel value calculation
   12–13
explosives–rock interaction   185

f
fire hazard   91
Fischer sub-sieve-size analysis   78
fly ash   145
free radical scavenging   171
freezing and thawing test   141–142
fudge point   140
full face tunneling   191
fume characteristics   52–55
functional safety, during AN explosive
   manufacture   163–165
– AN/FO   165–166
– electrostatic ignition   167–168
– explosion hazards in equipment   172
– – associated with pumping of explosives
   172–175
– – possible hazards during packing
   175–176
– explosion suppression technology
   171–172
– lightning protection   168
– runaway reactions   168–170
– slurries and emulsions   166–167
– venting as means of protection   170–171

g
gallery tests   99–100
gap test   22–23, 87
gelatins   35
gel rheology   89
gel rigidity test   89
general-purpose explosives   2

general purpose small-diameter explosives
   (GPSD)   105
– composition of   *148*
– design criteria and composition   105–106
– emulsion explosives   147–149, *150*
granulometry   81
grease   114, 129
guar flour   96
guar gums (GGs)   61, 65, 67, 72, 92–93
– application in water gels and slurries
   93–94
– cross-linking   95–96
– hydration mechanism   96–98
– typical specification used in water gels
   94–95

h
heat of detonation. See heat of reaction
heat of reaction   11–12
heavy ammonium nitrate/fuel oil (HANFO)
   52, 151–152
high-density prills   37
high explosives   5
history   31
hollow particles   126–127
horizontal batch mixing   39–40
hot spots   7, 116
hydrophilic–lipophilic balance (HLB)   130,
   134, 139
– use in emulsification   133–135
– – effect of factors on stability   135–138
hydrostatic pressure effect   90, 143

i
ideal detonation   9
ideal diameter   9
impact test   27
inorganic sensitizers   106–107
in-process and finished product checks
– mixing   85–86
– oxidizer blend composition   84–85
– packing   86
– solid ingredients   85

l
large diameter explosives   2
large-diameter packaged product (water gels)
   60
lecithin   129
low density ammonium nitrate (LDAN)   50
low density porous prilled ammonium nitrate
   (LDPPAN)   39
lower density prills   37
low explosives   5

m

manganin gauges   188
manufacture, of ammonium nitrate   35–36
– prilled ammonium nitrate   36–38
manufacture process, for emulsion explosives
– batch process   119
– critical equipment for production   121–122
– fully continuous process   120
– packaging equipment   122
– pumps   122
– semicontinuous operation   119–120
medium diameter explosives   1
microballoons   109–110, 139–140, 143, 147, 149
microbubbles   59, 89, 103
microcrystalline wax   123, 125
microemulsions   197
microspheres   109
MMAN   63, 103, 105, 107–108, 138
molasses   55

n

nanotechnology   197
nauta mixers   40
nitric acid   138
nitroglycerine (NG)   1, 15, 78, 100, 102, 104–105, 110, 157, 163, 169, 179, 192, 195
nondeflagration   100
nonideal detonation   9
nonionic surfactants   133
non-Newtonian character   136

o

occlusion temperature   125
oil absorbency and porosity   42
opencast mines   196
optimum scale depth   26
organic sensitizers   107–108
oxidizer blend   140
oxidizer salts, in emulsion   137–138
oxygen balance (OB)   61–62, 141, 169
– and fuel value calculation   12–13
– importance of   16–17

p

packaged-large-diameter emulsion explosives   153–154
paraffin wax   123
partition, of energy   11
pentaerythritol tetranitrate (PETN)   1, 169, 183
perchlorates   63
performance tests   86
perlite   127, 145

permissible explosives   2
– deflagration tests   100–101
– design criteria   98–99
– permissibility tests   99–100
– permissible water gel strength   104–105
– toxic fumes and typical formulation   104
– water gels behavior in permissibility test   101–103
permissible water gel strength   104–105
phase transition, and importance in explosives   33–34
phosphatide compounds   129
physical and chemical properties, of ammonium nitrate
– basic data   32
– decomposition chemistry   32–33
– phase transition and importance in explosives   33–34
piezoelectric gauges   188
plate dent test   24
polyatomic alcohol types   133
polyethylene   124
polyglycols polyoxyethylene types   133
polyisobutylene   124
polyisobutylenesuccinic anhydride (PIBSA)   132
polypropylene   124
precursor shock wave (PSW)   145
prills   15–16, 33, 36–38, 42, 71, 165
priming and boostering   47–49
process hazards, during slurries and water gels manufacture   91
– health hazard   92
– slippery floor   92
promoters, emulsion   128
pumpable explosives (bulk explosives)   3

q

quality checks
– aluminum in water gels and slurry explosives   74–84
– gel condition evaluation   89–90
– hydrostatic pressure effect   90
– in-process and finished product checks   84–86
– new formulation development   73–74
– performance tests   86
– process audit   140–141
– raw materials   71–73, 139–140
– safety tests   87
– special tests   141–142
– storage tests   87–89
– waterproofness test   90
quenching   171

r

reaction chemistry
– heat of reaction   11–12
– hierarchy rules   12
– oxygen balance and fuel value calculation
   12–13
reactivity
– definition of   123
– of AL powder   83–84
research and development (R&D)   157–158,
   195
– areas of interest   158–159
– development work and upscaling   159–161
– management   161–162
rheology   115–116
rifle bullet test   104
runaway reactions   168–170

s

safety and stability characteristics assessment
– accelerated hot storage (ageing test)   27–28
– cold temperature storage test   28
– impact test   27
– thermal stability tests using DTA and TGA
   procedures   28
– torpedo friction test   27
safety tests
– COD   87
– gap test   87
sensitizers
– air/gas/synthetic bubble sensitizers
   108–111
– inorganic   106–107
– organic   107–108
shear rate   116
shockwave, supersonic   8
slurries and water gels   59
– design   59–60
– – basic composition and process   65–66
– – bulk delivery product   61
– – ingredients list   60
– – large-diameter packaged product (water
   gels)   60
– – oxygen balance   61–62
– – small-diameter, cap-sensitive water gels
   60–61
– – thumb rules for design   62–63
– – water role   63–65
– general purpose small-diameter explosives
   (GPSD)   105
– – design criteria and composition   105–106
– guar gums   92–93
– – application in water gels and slurries
   93–94

– – cross-linking   95–96
– – hydration mechanism   96–98
– – typical specification used in water gels
   94–95
– permissible explosives
– – deflagration tests   100–101
– – design criteria   98–99
– – permissibility tests   99–100
– – permissible water gel strength   104–105
– – toxic fumes and typical formulation   104
– – water gels behavior in permissibility test
   101–103
– process hazards   91
– – health hazard   92
– – slippery floor   92
– process technology
– – batch process   66–68
– – continuous process   68
– – packaging systems   68–70
– quality checks
– – aluminum in water gels and slurry
      explosives   74–84
– – gel condition evaluation   89–90
– – hydrostatic pressure effect   90
– – in-process and finished product checks
      84–86
– – new formulation development   73–74
– – performance tests   86
– – raw materials   71–73
– – safety tests   87
– – storage tests   87–89
– – waterproofness test   90
– sensitizers
– – air/gas/synthetic bubble sensitizers
      108–111
– – inorganic   106–107
– – organic   107–108
small-diameter
– cap-sensitive water gels   60–61
– explosives   1
sorbitan monoleate (SMO)   131
– specification   *132*
soya lecithin   129
spallation   183
stability test   82–83
stabilizers, emulsion   128–129
stearates   132
stearic acid   132
stemming and confinement   184
storage tests   87–89
suppression, of explosion   171–172
surfactant   132–133

t

temperature cycling effect resistance
   44
tests   17–18
– aquarium   23
– ballistic mortar   18–19
– crater   26
– cylinder (crushing test)   24
– double pipe   23–24
– gap, and continuity of detonation test
   22–23
– list   18
– plate dent   24
– Trauzl lead block   19–20
– underwater   24–26
– velocity of detonation   20–22
thermal stability tests using DTA and TGA
   procedures   28
torpedo friction test   27
toxic fumes and typical formulation
   104
Trauzl lead block test   19–20
tray packer   69
trinitrotoluene (TNT)   1, 59
tunnel blasting   191

u

underwater test   24–26
USBM test for deflagration   101

v

velocity of detonation (VOD)   3, 5, 9, 10, 16,
   59, 73, 76, 86, 102, 104, 127, 136, 143, 146,
   157, 184, 188, 191
– and effective priming   47–49
– of explosives confined and unconfined   *144*
– test   20–22
– test setup for   *23*
venting, as means of protection   170–171
vermiculite   127, 145

w

water-covering value   81
water gels behavior in permissibility test
   101–103
water in oil (W/O) emulsion   114, 118, 123,
   130–132, 134–135, 137, 142
waterproofness test   90
water resistance   142
– test   51
wetting   172